Discrete Problems in Nature Inspired Algorithms

Discrete Problems in Nature Inspired Algorithms

Anupam Shukla

Ritu Tiwari

CRC Press
Taylor & Francis Group
Boca Raton London New York

CRC Press is an imprint of the
Taylor & Francis Group, an **informa** business

CRC Press
Taylor & Francis Group
6000 Broken Sound Parkway NW, Suite 300
Boca Raton, FL 33487-2742

First issued in paperback 2020

© 2018 by Taylor & Francis Group, LLC
CRC Press is an imprint of Taylor & Francis Group, an Informa business

No claim to original U.S. Government works

ISBN-13: 978-0-367-57237-2 (pbk)
ISBN-13: 978-1-138-19606-3 (hbk)

Visit the Taylor & Francis Web site at
http://www.taylorandfrancis.com

and the CRC Press Web site at
http://www.crcpress.com

Contents

Foreword I

Nature-inspired algorithms are currently gaining increasing popularity for their widespread applications in diverse domains of scientific and engineering optimization problems. There exists a vast literature on nature-inspired algorithms, spread across scientific journals, conference proceedings, and so on. Unfortunately, I did not come across any book covering all aspects of nature-inspired algorithms. This book by Professor Anupam Shukla and Dr. Ritu Tiwari fills this void. It covers almost all the aspects of nature-inspired algorithms in a single, precise, and resourceful volume.

A lot of promising nature-inspired algorithms have emerged over the past two decades. Particle swarm optimization, ant colony optimization, bat algorithm, artificial bee colony algorithm, bacteria foraging optimization algorithm, shuffled frog-leaping algorithm, cuckoo search algorithm, invasive weed optimization algorithm, and flower pollination algorithm are just a few to mention in this regard. At this point, there is a need to educate people about the principles, concepts, applications, issues, and solutions related to the use of nature-inspired algorithms.

I am sure that this book will be of great contribution toward the same and would further carve a deep mark in the nature-inspired algorithms literature for the reference of enthusiasts. It is unique for its content, readability, and above all the presentation style. This book includes both the theoretical foundations along with a variety of applications that further help in understanding the different nature-inspired algorithms. It is always considered to be good to have a practical component attached to the courses whose performance ultimately reflects the overall understanding of the students. Ample coverage of applications, issues, and perspectives, makes this book rich and diverse. This book will be very useful for the students, academicians, and researchers studying/working in the field related to nature-inspired algorithms. It would be equally useful to researchers migrating from mathematics and other disciplines to computer science.

Amit Konar
Jadavpur University, Kolkata

Foreword II

This book provides a systematic introduction to most of the popular nature-inspired algorithms for optimization, including particle swarm optimization, ant colony optimization, bat algorithm, artificial bee colony algorithm, shuffled frog-leaping algorithm, invasive weed optimization algorithm, flower pollination algorithm, and so on.

Nature-inspired computation is one of the most promising areas of research and has a lot of scope in the near future as well. Nature-inspired algorithms have very diverse application areas, which are growing steadily. Nature-inspired algorithms have in fact become the most widely used optimization algorithms. This book covers the theoretical background including extensive literature review of the nature-inspired algorithms and the practical implementations of these algorithms for solving various discrete optimization problems.

This book will definitely be useful as a reference book on nature-inspired algorithms for graduates, postgraduates, doctoral students, researchers, and even faculties in computer science, engineering, and natural sciences.

P.N. Suganthan
Nanyang Technological University

Foreword III

Combinatorial optimization forms an important area under theoretical computer science and applied mathematics with several industrial applications and it aims at identifying an optimal solution from a finite set of solutions. The solutions are normally discrete or can be formed into discrete structures. In most of such problems, the exhaustive search becomes intractable, necessitating various forms of intelligent search and optimization. This book—*Discrete Problems in Nature Inspired Algorithms*, authored by Professor Anupam Shukla and Dr. Ritu Tiwari—comprises a great treatise on solving the usually NP-complete combinatorial optimization problems by using various nature-inspired metaheuristic algorithms. Starting with a lucid introduction to various combinatorial optimization scenarios, the author has aptly guided the reader through a comprehensive journey of the development of the metaheuristics, their qualitative advantages and downsides, and applicability to various related problem domains. The exposure of the material is lucid. Quite complicated concepts are presented in a clear and convincing way, which can be attributed to the expertise of the author.

Finally, I must conclude that this is a very timely volume with a well-compiled exposure of the cutting edge research on and with nature-inspired discrete optimization algorithms. I am quite sure that the researchers and practitioners from the related fields will be immensely benefitted from this book.

Swagatam Das
Indian Statistical Institute, Kolkata

Preface

Nature has been a constant source of inspiration for scientists since centuries. Various laws and theories have been formulated by observing the different phenomena that occur in nature. *Nature-inspired algorithms* is an area inspired by natural processes; in the recent past, many algorithms have been developed taking inspiration from certain biological, geographical, and chemical processes that take place in the natural world. Nature-inspired algorithms have the immense potential for optimizing complex real-world problems and have been applied in various areas. In the past two decades, plethora of new and promising nature-inspired algorithms has been introduced. The highly multidisciplinary nature of the field further has attracted a number of researchers. There is a mammoth of literature available in the field of nature-inspired algorithms. The current literature covers the theoretic concepts, but the practical applications might still be very difficult for students to comprehend and implement. Many students find it difficult to visualize the application of nature-inspired algorithms from theoretical text.

This book not only helps them in understanding these algorithms but also exposes them to the current developments over various related fields. This empowers them to pursue research in these areas and contribute toward research and development at large. The main purpose of this book is to explain the nature-inspired algorithms theoretically and in terms of their practical applications to real-world problems to the readers. This will enable them to get an in-depth, practical exposure to the methodology of the application of the nature-inspired algorithms in various situations.

This book can also be used as a standard text or reference book for courses related to nature-inspired algorithms, evolutionary algorithms, and so on. This book may be referred for the purpose of *elementary* and *advanced* studies as it introduces the basic concepts of nature-inspired computing and then also discusses some state-of-the-art variants of those basic algorithms and then finally explains their practical examples. This book also incorporates some of the recent developments over various areas.

Salient Features

- Introduction to the world of nature-inspired computing
- Foundations of the basic nature-inspired algorithms
- Detailed description of the problem and solution
- Information on recent developments in the various nature-inspired algorithms
- Interdisciplinary applications of the nature-inspired algorithms

Intended Readers

- Students of undergraduate, postgraduate, doctorate, and postdoctorate levels can refer this book as a text or reference book for nature-inspired algorithms.
- Researchers can refer this book to obtain a good insight into the recent developments in the field of nature-inspired algorithms.

Origin of This Book

This book is a result of the continuous research done over the time by the authors. Much of the work presented is done by the authors themselves. Databases to the various problems have either been self-generated by the authors or used from the public database repositories. At some places, the authors have also included a comprehensive discussion of the problems and the solutions present in the literature. This is a result of the efforts put in by various students of the ABV-Indian Institute of Information Technology and Management, Gwalior, India, in forms of projects: B.Tech, M.Tech, and PhD theses.

Acknowledgments

This book is our yet another step into the world of authored books. It began as a dream to use our past research as a base to create a landmark into the world of nature-inspired computing. It is natural that the dream required a lot of encouragement, guidance, and support that were provided by numerous people in various phases of this book. The authors acknowledge all those people who made this possible.

The authors thank Professor S.G. Deshmukh, director, ABV-IIITM, Gwalior, India, for extending all sorts of help and support during the course of writing this book. This book would not have been initiated without his encouragement and support.

The authors thank Professor. Amit Konar, Professor P.N. Suganthan, and Professor Swagatam Das for writing the foreword for this book. Their precious words have added a greater charm to this book. The authors further express their thanks to Dr. A.S. Zadgaonkar, ex-vice-chancellor, Dr. C.V. Raman University, Bilaspur, India, for the guidance and motivation that he has bestowed during the entire course of work with the authors.

This book would have never been possible without the pioneering work carried out by the coresearchers associated with the authors. The authors also thank the student fraternity of the Institute for taking deep interest into the courses titled "Special Topics in Intelligent Systems" and "Soft Computing." Their interest to learn beyond the classroom program has been the key motivation for this book. The authors are highly indebted to all the students who undertook challenging problems as a part of their thesis and came up with excellent solutions and experimental results. The authors thank their PhD students Sanjeev Sharma, Apoorva Mishra, and Saumil Maheshwari for their constant support during each phase of this book. The authors thank their postgraduation students Prashant Shrivastava, Chiranjib Sur, Annupriya, Riya Naval, Alok Sharma, Prashant Pandey, Harshita Lalwani, and others for the quality results they produced during their course of work with the authors. The authors further thank Pritesh Tiwari for helping in editing some parts of the manuscript.

The first author thanks his son Apurv Shukla for his support during the course of writing this book. The second author thanks her mother Parvati Tiwari and her late aunt Dr. Asha Shukla for motivation and blessings for writing this book.

This book required a lot of work by numerous people around the globe. The authors thank the entire staff of Taylor & Francis Group and all the associated units who have worked over different phases of this book. The authors thank all the people who have been associated with and have contributed to this book. Above all, we all thank the Almighty for his constant blessings and love. Without Him, this project would not have been possible.

Authors

Prof. Anupam Shukla is currently a professor in the Department of Information and Communication Technology (ICT) at ABV-Indian Institute of Information Technology and Management (ABV-IIITM), Gwalior, India. He has 29 years of administrative, research, and teaching experience. He is globally renowned for his research on artificial intelligence and robotics, which has won him several academic accolades and resulted in collaborations with academicians across the world.

He has been awarded "Distinguished Professor" by Computer Society of India, Mumbai, 2017; "Dewang Mehta National Education Award" for best Professor, 2016; "Professor Rais Ahmed Memorial Award," Acoustical Society of India in 2015; Engineering Professor of the Month, Higher Education Review, India, 2014; "Best Paper Award," Institute of Engineers, India, 2006; University Gold Medal (Valedictorian) from Jadavpur University, Kolkata, India, 1998, for scoring highest marks at master's level; and "Young Scientist Award," Madhya Pradesh Council of Science and Technology, Bhopal, 1995.

He is the author of three patents, three books titled: (1) *Real Life Applications of Soft Computing*, Taylor & Francis; (2) *Intelligent Planning for Mobile Robotics: Algorithmic Approaches*, IGI Global; and (3) *Towards Hybrid and Adaptive Computing: A Perspective*, Springer–Verlag Publishers, editor of three books published by IGI Global Press, United States, and mentor of 17 doctorate and 106 postgraduate theses. He has 174 peer-reviewed publications and is a reviewer of various international journals, including *ACM Computing Review, IEEE Transactions on Information Technology in Biomedicine*, and *Elsevier Neurocomputing journal*. He has successfully completed 13 government-sponsored projects, aimed at developing IT applications for Indian farmers, skill development of the Indian youth and infrastructure development at parent institutes, and so on. He is currently a nominated expert member of the accreditation committee of National Board of Accreditation, India. At NIT Raipur, he was the founding Head of the Department (HOD) of departments of biomedical engineering and biotechnology engineering and the founding coordinator of the Chhattisgarh–IIT Kanpur Knowledge Sharing Program, a first of its kind e-learning program in India.

Dr. Ritu Tiwari is an associate professor (Department of Information and Communications Technology) at ABV-IIITM, Gwalior, India. She has 15 years of teaching and research experience, which includes 10 years of post PhD teaching and research experience. Her field of research includes robotics, artificial intelligence, soft computing, and applications (biometrics, biomedical, and prediction).

She has two patents in her name and has authored three books titled: (1) *Real Life Applications of Soft Computing*, Taylor & Francis; (2) *Intelligent Planning for Mobile Robotics: Algorithmic Approaches*, IGI Global; and (3) *Towards Hybrid and Adaptive Computing: A Perspective*, Springer–Verlag Publishers. She has also edited two books in the area of biomedical engineering from IGI Global. She has supervised 7 PhD students and more than 100 master's students and has published 104 research papers in various national and international journals/conferences.

She has received "Young Scientist Award" from Chhattisgarh Council of Science & Technology, India, in the year 2006. She has also received gold medal in her postgraduation from National Institute of Technology (NIT), Raipur, India. She has completed 10 prestigious research projects sponsored by Department of science and technology (DST) and Department of Information Technology (DIT), Government of India. She is currently involved with the Government of India and is working on three sponsored research projects. She is a reviewer of various international journals, including *ACM Computing Review, IEEE Transactions on Information Technology in Biomedicine, Elsevier Journal of Biomedical Informatics,* and *Elsevier Neurocomputing journal.*

1

Introduction to Optimization Problems

1.1 Introduction

In the present world, we are obsessed with solving a huge number of problems. With the progress of science and technology, humans have been able to tame some of them [1]. Due to advancements in the field of automation, and knowing the fact that natural resources are on the verge of exhaustion, scientists, today, are paying a lot of attention on the low resource (regarding power, time, and occupancy) factor, which can be achieved through optimization. Optimization can be achieved at many levels.

The discovery of nature-inspired computing and its allied optimization branches has revolutionized computation and automation, the future of which lies in the hands of machine learning and self-adaptive systems [2–4].

One might wonder, why adaptive learning systems? In a real-life environment, every system is associated with some input and output mappings in which the inputs act as the main decisive factor for the outputs. However, unless the system is carefully designed, the range associated with the input and output data variables is highly unpredictable and nonstationary, and there exist a large number of samples. In such a scenario, it becomes a challenge for supervised learning to establish a nonlinear model, which can produce correct judgment through the efficient processing of data and intelligent decision-making. However, there always remain chances of deviation due to irregularities in the datasets. Hence, there is a requirement for the system to adapt to the appearance of a new data and the combination of situations.

This adaption is handled in various ways in various situations and it depends on the mathematics of the algorithms. In this chapter, we will introduce techniques for such kinds of adaption that are involved in nature-inspired metal-heuristics [5] on many problems.

This work can be marked as an introduction to a bunch of bio-inspired computation techniques, which can be readily applied to the graph-based and discrete optimization problems. Before moving further, the term *discrete optimization problems* should be explained and illustrated so that you can have a clear idea of what kind or class of problems we are referring to. Consider an equation of any straight line of the form $y = mx + c$. For any value of $x \in R$, where R is the set of Real numbers, the value of y is valid and acceptable and this is a simple continuous domain problem. There is no minimum or maximum value that can be achieved due to the nonconvergence attitude of this equation. On the other hand, an equation of the form $y = |x_1^2| + |x_2^2|$ has a minimum value at zero and x_1, x_2 both can have any value in R. What if we say that the equation $y = |x_1^2| + |x_2^2|$ also has a minimum value at zero but x_1, x_2 can only take integer values lying within Z, where Z is the set of Integers.

This is what is referred to as the fundamental discrete optimization problem in which the variable parameters take only integer values. However, this is not what we are looking for. We are going for a generalized discrete optimization problem set, which will have the above-mentioned kind of problems as special cases and we would not have to reformulate our equations.

Now, optimization may mean different things for different types of problems. For minimization as optimization of a function f, we have $f(x) \leq f(a)$ where $\forall_x \in S$ and $S \subseteq B$ where S denotes the allowed representations for x and B is the constraint for the variable, which can be a domain or range or an integer value depending on the problem. Here, $a \in S$ and $f(a) = \text{MIN}(f(x) : \forall_x \in S \subseteq B$ and is called the optimized solution or the global solution.

1.1.1 Artificial Intelligence

Artificial intelligence is defined as the capability of perfect sensing and intelligent decision-making of a system, which facilitates complex jobs and is worthy of usage for humankind and works under the influence of an external environment (virtual or real). Artificial intelligence can also be defined as the capability of the performance of the system, which cannot be achieved through a fixed sequence of instructions. Systems based on a fixed sequence of instructions would be biased, and an optimal decision would not be possible [6].

In other words, what artificial intelligence does is the same as done by a mechanical machine and that is it *makes life easy*. Similar to machines, it helps in performing a very complex task in the blink of an eye or with the press of a button. The same is done by a *simple computer program*, but can we call it an artificial intelligence system? No, we cannot! And there are several reasons for it. A *simple computer program* is just performing or executing some instructions and, at any point in time, it does not take any decisions, and it is not bothered whether the output generated is desired or not. As we are talking about taking decisions, it is obvious that more than one outcome or combinations of outcomes is involved, but the system will seek out the best or the optimized one (if it can, else it will let one decision go and later find out whether it was the best or if it is the best among all) for the given situation depending on some evaluation criteria. So, an intelligent system must be able to make complex judgments consisting of multiple parameters and situations taking care of the constraints and rules. In modern times, intelligent systems are adaptive and have the capability of learning from experience [6].

Consider the example of any search operation such as playing chess on a computer. The feedback from the users (i.e., the move they make) acts as an influence for the system to reproduce the optimal decision, which will make the game more interesting. Had the system taken any random step, its reaction would not have been considered intelligent, and the result of such a decision would have been a matter of probability and chance. However, sometimes randomness appears to be more useful than more deterministic approaches, especially when the combination of decisions is quite large and consequently the cost (both regarding the computational complexity and time) of estimation of fitness of such a decision is also huge. The purpose of adoption of random behavior in many decision-making functions is to reduce the dimension or the number of possible combinations for all the entities that are involved in the system. Another example is the search engine that had revolutionized the ways of information storage and its retrieval capability when the quantity of such information had grown boundless, and there was a severe requirement for search engines capable of procurement and filtering. What was mainly done was the incorporation of artificial intelligence into the query-making system, which sorted the information of a particular regime according to the importance, and this one kind of ranking system helped in the quick retrieval of relevant information much quicker than expected. However, the mode, criteria, conditions, and implementation of such kind of intelligence-driven engines differ from organization-to-organization, and that is why for particular word search different search engines hit on different sequences of web pages. Another issue that needs to be mentioned is that these search engines are also monitoring systems, which take account of the statistics of usage and frequency of visit at a website and, accordingly, evaluate and maintain the importance of the information of that site. It is worthy of mentioning that with usage, experience, and time the intelligence of the search engines tends to get changed and enhanced.

1.1.2 Soft Computing

Soft computing is another important sister branch of artificial intelligence in which the system starts acknowledging, accepting, and becomes capable of processing values from a continuous range, instead of grouped implicated values. The main conception arose from fuzzy logic in which the system design is flexible, and the data ranges are subdivided into more overlapping classifications and subclassifications. The implications have been proven to generate more accurate decisions and desirable results. In real world, sometimes it is very difficult to define or clearly demarcate ranges for a particular event such as temperature (hot or cold), height (tall or short), and turn (sharp or simple). The values of these events are relative, and numerical ranges are unclear. We do not know the numerical value for *hot temperature*. So, these events need more classifications before making a system intelligent enough to handle such range of values and provide proper decisions. In contrast to the hard rules what *soft computation* tried to develop is flexibility and rule-free classifications. As in the case of height if greater than a threshold is *tall* else *short*, what about the height that is very near to the threshold? Should it be classified as short or tall?

Another problem set that helped in the evolution of soft computing is the calculation of maxima and minima for the discontinuous functions. Differential calculus has been the key to finding the maximum and minimum for functions for years, but a function is only differential when it is continuous at that range. So, for the discontinuous functions, differential calculus is not valid, and this paved the way to the so vast field of nature-inspired computation and soft computing. Apart from fuzzy logic, two other important members of soft computation are neural network and genetic algorithm (GA), which also share their space with nature-inspired computation (as they are derived from the natural phenomenon). They are widely used for numerical optimization of several discontinuous mathematical functions. GA is discussed in detail in Chapter 3.

1.1.3 Intelligent Systems

Intelligent systems are the system, which are provided with an artificial intelligence and are capable of processing data and take decisions intelligently. Unlike expert systems [7], they can handle unknown environments and unknown events that are based on their decision-making capability. The main feature of this kind of system is that they can or may hover over all the possible solutions before making any decision. Playing chess on a computer is an example of an intelligent system. Before going for a turn, it will evaluate all the possible moves and then go for the best one. Intelligent systems are meant for external feedback of some limited range or number of inputs or inputs from a static environment.

1.1.4 Expert Systems

An expert system is one kind of intelligent system, which is more likely an intelligent pattern-matching system capable of making decisions based on the data that are fed as a part of the training. The expertise, intelligence, and decision-making are done on the basis of feed data only. Intelligence lies only for the known environment or the set of known events. A typical example can be a biometric system. An expert system is subjected to an unlimited number of external feedback as inputs, out of which it can only process correctly only the known ones. As for identification of a person, there are several characteristics that are unique. However, for big databases, a collection of such characters can be used as unique identification characteristics. The same is true when there are several features overlapping with the other entity in the same database. Some of the features are not adequate for describing the entity and in these kinds of situations the most overlapping features are regarded as noise factors. The relativity or correlation of noise depends on the number of entities it can represent. So, in an expert system, the learning models extract or rather learn more from the unique features and less from the other subset of features.

Theoretically, imagine a hypothetical situation where a student has to learn lessons. The first lesson is being learned sincerely rather exclusively, but if there are common features between kth and $(k + 1)$th lesson, then it is advisable to learn the extra features of both of them and only learn the common features from only one of them. So, basically, in a mathematical model the standing difference lies only in the unique features of both of them and this difference is then utilized as a feedback error for manipulation of the learning model through self-adjustable techniques such as back-propagation algorithm and so on.

Now the question is "What will be an intrusion detection system?" Is it an intelligent system or an expert system? An intrusion detection system monitors the behavior of a system and takes a decision based on the analysis of the processed data, but at the same time, it tries to linkup with the experience and previously collected knowledge of anomaly detection. It is very difficult to draw a strict classification margin between the systems and its types. So, intelligent systems rely on logic and expert systems rely on training, and there are data associated with them in common. In some sense intelligence of the expert system is *dumb intelligence*. However, the input data for both expert systems and intelligent systems have bounds and limitations, and decision-making for unknown data is characterized mainly by machine-learning intelligence.

1.1.5 Inference Systems

Intelligence or decision-making in inference systems is based on conclusive decisions made out of rational thinking and not on logic. But should we call it nonlogic decision-making? In an inference system, the control statements are cognitive statements, which may or may not be represented mathematically, and thus are very difficult to process for computation. The control statements arise out of rational thinking, and the basic meaning of the statement is not important, but its implications are important for articulating meaningful relationships.

Consider the two mathematical statements. One is $x > 10$ and another $y > x$. Combining the above-mentioned two statements we can conclude that $y > 10$. If the statements would have been like *x is greater than* 10 and *y is greater than x*, then also we could have concluded as *y is greater than* 10 but only through the use of mathematical logic. But what if the statements are not mathematical? For example, *The sun rises from The East* and *Mohan is looking at the rising sun*. The conclusion could have been drawn like *Mohan is looking toward east* (*unless he is using a mirror or watching some video*), but how? There is no mathematical logic behind it, and the conclusion is correct. An inference system plays the role of drawing decision and conclusions from this kind of statements better known as on-logic decision-making. However, there are challenges for inference system for comprehension and handling many nonconclusive sentences.

1.1.6 Machine Learning

Machine learning is another high-level abstraction-based learning phenomenon in which the algorithm gradually explores the environment and its objects and learns from it. Initial moves are random or user defined but later they are updated with the progress of exploration and are also on the basis of monitoring with the learned experience. All the learning process and its storage are updated in the form of state variables and numerical values that indicate some implications. Machine learning is an iterative process, and the system gets better with experience. There are several algorithms for machine learning, which are beyond the scope of this book. There is the strategy of analysis and learning, but there is no specific rule for data representation and learning models in machine learning and each one of them have their own advantages over the other. In some cases one technique proves to be better than the others, whereas in some different situation, the other technique might outperform it. This concept is known as no free lunch theorem [8].

Consider the YouTube search engine; when you search for any video, it not only provides all its versions but also some other video, which is somehow relevant to the searched one and most of the times they are pertinent and useful mainly during the less specific generic word(s) search. How does a search engine know what was I looking for and, more importantly, my taste? There is nothing called magic, and there is hardly any person who is going to spend so much time linking up the similar videos among huge data storage. There are machine-learning algorithms running behind and from the pattern of people watching the videos, the algorithm learns and, accordingly, set up links that are based on priority and frequency of visits, and this statistical monitoring and mathematical framework are acting as a suggestion for other viewers. The machine-learning algorithmic process of refinement of suggestions is getting better day by day. Initially, when there was no learned suggestion model, the suggestion links were random or based on cluster bounded by region, state, or country. Even now the filtering and suggestion based on the country can be manually set in YouTube. Filter bounds are there to make the suggestions and filtering process less complex, and thus the diameter of the graph representing links between a video and its suggestions remains low.

1.1.7 Adaptive Learning

Adaptive learning is yet another process of learning-based models for optimization and control-based systems. During the search process for system optimization, there is a requirement for adaptively learning and responding to the requirement of step variation and identifying the regions where there is more possibility of finding optimal or near optimal solutions occurs. This kind of response is obtained through a feedback generated by a mathematical representation whose response is adaptive and based on the parameters and evaluation generated. This learning model helps us to quick convergence rate and local search and prevents unnecessary swing of space search domain. They are mainly used in the control system and mathematical optimization. In the control system, the strength of control features needs to be adjusted according to the ongoing state of the system and thus, need adaptive mathematical modeling. For example, in a power plant, the inputs of fuel need to be regulated if the temperature or pressure or both are high. However, for numerical optimization of mathematical equations, the step-wise variation must direct toward the place where there always occurs improvement of fitness and when the peak near the step-wise variations is small. However, there is another important requirement for adaptive learning for the neural network during the training phase in which the parameters of the intelligent system (consisting of neural network models) need to be manipulated or tuned so that the error between the original output and the expected output gradually disappears. This kind of system and adaptive control is very helpful in biometric systems, pattern matching-based expert systems, and numerous applications.

1.1.8 Related Work Done

1.1.8.1 Heuristics

Heuristics can be defined as a cognitive search-based algorithm for quick optimization or near-optimized solution, which cannot get solved with the conventional mathematical tools and exhaustive search techniques. In other words, such systems or problems cannot be represented by continuous gradient-based mathematical representation or may have multiple local peaks for minima or maxima. Such cases are similar to nonconvex problems such as multimodal problems, multilocal optima problems, and discontinuous equation-based problems, and are difficult to be solved analytically and sometimes infeasible and impractical. The solutions are measured with the help of an evaluation function or called heuristic function, which is convenient in many cases, but in the majority of the systems and their modeling (such as real life and real time), it is very difficult to achieve and formulate that function. The convenience in the determination of this numerical evaluation is achieved through abstraction and efficient data processing. Examples of heuristic searches are dominated by only random exploration or by the combination of exploration and exploitation. Typical and simple examples of heuristics are educated or wild guess, an intuitive judgment-based guess, sixth sense, and so on.

1.1.8.2 Meta-Heuristics

Meta-heuristics is a semirandomized search algorithm in which the optimal solution of an NP-Hard problem is derived out through the cooperation of agents with the help of deterministic and stochastic operations, which constitutes the algorithm. The joint acknowledgment of deterministic and stochastic operations engages the algorithm in both exploration and exploitation simultaneously. The only deterministic process would not have been able to produce the optimized solutions due to limitations in combination determination and the high cost of the evaluation. The introduction of randomness in parameter generation or direction of movement or acceptance of position or event is the key to success for many meta-heuristic algorithms. The main reason for such success is opportunistic combination formations and escaping from the local minima similar to a situation for the unexplored areas and may be toward global optimization. This random factor has several forms in different aspects and has different operations in different algorithms. A typical example is better known as a mutation factor in evolutionary algorithms such as GA and differential evolution. Every nature-inspired computation can be regarded as a meta-heuristics in some form or the other. If the random features are not present in algorithms, then they would behave similar to a greedy algorithm or deterministic algorithms such as Prim's algorithm and Dijkstra algorithm. Two salient features of meta-heuristic algorithms are intensification and at the same time some diversification. It is another way of saying exploration and exploitation. Diversification means a deviation of mean that is away from the best results or global optimization region. Intensification is another way of reinventing the adjacency of the global best region for better positions. So, in the intensification phase, agents hover around the global best search space and try to detect better solution around, if possible, through variation of parameters. While in the diversification phase, it is ensured that the agents do not leave out of the unexplored search environment. An artificial bee colony (ABC) that consists of scouts and workers is a typical example of the combination of intensification and diversification. ABC is described in Chapter 4. So, briefly what we mean is "The specific objectives of developing modern meta-heuristic algorithms are to solve faster, to solve complex problems, and to obtain more robust methods." Meta-heuristic draw their potential and way of solving problems through the source of inspiration such as any natural phenomenon and in the 2 to 16 chapters we are going to discuss various kinds of such meta-heuristics and their wide range of application.

1.1.8.3 Hyper-Heuristics

Hyper-heuristics is an automated search scheme-based algorithm in which the operations are performed based on the previously learned experience. In other words, it is a cooperative approach of search and machine learning in which the search is driven by an experience that is gathered by learning models. In hyper-heuristics, the search is accompanied by an intelligent decision-making of the choice of parametric variations and operations in the search space. If the best solution is near the peak, in meta-heuristics we will move on exhaustively for new options, but in hyper-heuristics, the agents will only move if the better position is reached as the learned model only searches in the position where it has found better positions. Sometimes the main search process is accompanied by exhaustive local searches, which come into act only when the agents have reached a considerable stage. In particle swarm optimization (PSO), if the particle keeps on moving with the same speed then it is meta-heuristics, but if it lows down with the movement toward the peak, then it is hyper-heuristics. PSO is described in Chapter 2 and GA is described in Chapter 3. In addition, now at the best position, if we implement a GA-based PSO, then it is again hyper-heuristics, instead of just hovering around the search space randomly. The GA-based PSO hybrid has been proven to have a better convergence rate than the PSO itself.

1.1.8.4 Nature-Inspired Computation

Nature-inspired computation is the generalization of a bio-inspired computation and evolutionary computation where in each case the algorithms are derived from the natural phenomenon of the surrounding environment or nature and its habitats. The operations of the algorithms are a replica of what happens in nature and sometimes are mathematically modeled so that it can approximately represent such phenomenon. The nature-inspired algorithms have been very successful in the optimization of many NP-Hard problems, combinatorial optimization problems [9], and many mathematical representations. In the subsequent we will be discussing the application of the nature-inspired algorithms for graph-based and discrete optimization problems and will try to uphold the salient features of the algorithm.

1.1.8.4.1 *Adaptability in Nature-Inspired Computation*

Adaptability is the mode of tuning the various parameters of the algorithm for enhancing its performance and the use of adaptation varies from process-to-process and mainly helps in thorough local search, and to quick convergence. The adaptation also produces nonuniformity and sometimes scattered or random phenomenon and thus helps in more exploration in search space. Another characteristic of adaptation is the mathematical intelligence that can converge toward the peak values both in the case of minima and maxima. Mathematical intelligence helps in deciding direction toward optimum and also adaptively decides the step-size according to the progress of the algorithm. It is like your eyes are covered and you are looking for something. So until you reach the probable place, you will take big steps for quick coverage of the unnecessary regions and when you reach the probable place you will take small steps for thorough and enhanced local search. This phenomenon is illustrated elaborately in the subsequent chapters when the adaptive version of the Nature Inspired algorithms is considered. Another important feature of the adaptation is the ability of reconfiguration of the intelligent system with the help of error feedback similar to that in the control system. It is very difficult to design an intelligent system for a real-world problem because of the nonlinearity in every sector of data; and moreover, representing a system in mathematical form is very challenging. So, the researcher first defines an approximate nonlinear system to represent the intelligent system and then with iteration it tries to modify the system parameters adaptively, and thus after much such trial the nonlinear takes the form of an expert system. This is what happens in systems involving feedback systems such as control system design and neural networks.

1.1.8.4.2 *Learning in Nature-Inspired Computation*

Nature-inspired computation techniques are mostly adaptive in nature as most of the algorithms are operated in an unknown environment search space and it is considered that the agents or the system of agents have no information or experience of the environment. This is done to make the algorithm robust, generic, and ready for any situation. In addition, in dynamic environments, the change in network parameters should also be taken care of and in that scenario, the previous experience will be of no use or may be misleading. However, incorporation of learning methods and statistics of previous best results will make the algorithms stronger.

Constraint handling in nature-inspired computation: Algorithms, in general sense and mainly, take decisions dynamically through the process of analysis that are forward on the current state space and go ahead accordingly and simultaneously take care of the constraints easily during step-wise progress; they mainly make sure that the parameters respect the domain and the range of the entities.

Perhaps, in such cases, every time the algorithm is executed, it tends to provide the same result (set of intermediate and end results). However, in nature-inspired computation-based algorithms, decisions are taken probabilistically or based on a hybrid of stochastic and deterministic method or in a semideterministic way. Operations are operated independently and then the solution is validated and its fitness is determined. The mode of independent application of algorithmic operations and use of multiple solution agents make the algorithms to quickly converge to an optimum or near optimum, and hence the algorithms have gained so much popularity in the multidimensional optimization domain. As most of the operations related to these algorithms are processed independently, the constraints related to parameters or the mathematical equations are not taken care of during operation. Instead, the validity is opted during fitness evaluation. Obviously, if constraints are violated, the solution is either discarded or rolled back to the previous one.

1.1.8.5 Multiagent System

When a complete set of assignments is accomplished by the communicated or noncommunicated coordination of several agents (static or mobile), where each one of them performs their assigned task as part of their ability and utility and of the assignment, it is called multiagent system. Needless to mention that several agents can be assigned to do the same job but to maintain efficiency the whole coordinated system always tries to avoid redundancy and duplicity of the same event. However, multiagent-based simulation study has always been advantageous in which several agents are assigned to do the same task dependently or independently because it provides a visualization or rough estimation of what would have been the most efficient and profitable structure or combinations that are subjected to proper modeling of the system and its characteristics. However, when several agents coordinate and work together they save time, energy, enhance robustness, and also maintain least chances for failure. Complexity is also reduced for coordination, communication, and fault while handling the complexity of the system and intelligent decision-makings.

1.1.8.6 Multiagent Coordination

Multiagent coordination is the phenomenon by which a group of agents work simultaneously under the influence or reference of its neighboring agents or in some cases one or more agents may be selected as leader agents from a pool. The aim of such coordination is accomplishment of a part of the whole work. The motive of this kind of coordinated work thus lies in providing efficiency of the system with respect to time, energy, duplicity, computational power, the cost of decisions, and complexity of the overall system. The coordination of the agents in multiagent-based systems [10] is achieved through communication or sometimes virtual adhesion/cohesion phenomenon through the feedback of a subset of a fully automated data-collecting set of coordinated sensor systems. There are various types of examples of this kind of multiagent coordination, which are widely utilized and are different for different kind of applications. Like in swarm intelligence, the coordination is mathematically relative and not cohesive as seen in the case of flocking-based robot movement or coalition-based accomplishment of the task. However, in case of a multisensor system, they operate independently as an important part of the system. In the case of exploration, the multiagent coordination is based on separation (of regions), whereas in the case of multiagent coordination in path planning [11], it is a division of work. In the case of multitasking jobs such as lifting a box, there is a requirement that all the agents exert the same force on all direction. Hence, in this kind of situations, the coordination is related to real time and is controlled by real-time monitoring and error feedback on the system and the individuals, and also depends on the mathematical modeling of the system and the feedback control [12].

1.1.8.7 Multiagent Learning

Learning from experience and exploration in an agent-based coordinate system is known as multiagent learning. Each of the agents may work independently or under the influence and coordination of other co-agents. Learning in a multiagent system can be shared or unshared and depends on how similar the agents are and how far the learning of one agent is going to help the other. In a leader-based system, the data accumulated by the joint effort of the agents are centrally processed to derive some learning-based experience and the processed outcome is shared with the agents. Learning depends on some evaluation function such as reward function reinforcement learning-based structure, the best state-space analysis in Q-learning-based structure, and fitness function for pattern analysis for data sets in an unsupervised learning. Supervised learning is rare in agent-based systems; however, it is not new when the data patterns are known and limited to some fixed known kinds. The learning models become challenging in uncertain environments and nonstationary environments such as adaptive feedback systems. In multiagent learning, the agents are more influenced by what has been learned (individually or centrally shared by other agents) than the temporary online influence of the social mates. For example, let us assume a graph-based network where one node represents one agent, who needs to reach another node in an optimized manner. Now, Meta-heuristic algorithms such as ant colony optimization (ACO) will induce the agents through the route with maximum pheromone involvement, but that may not be the optimized route though most of the ant agents might have followed it, the greedy algorithm will follow the most probable link by deterministically analyzing the available links through local heuristics. However, if learned, the next best state space will gradually acquire more rewards through experience and the agents will follow the reward matrix. Now, the reward matrix can be thought of as some entry reward matrix in which each agent gradually acquires more rewards cumulatively with iteration. A multiagent learning had been a good research for some applications. For example, in multicamera-based fusion of images, the central unit must learn and understand that in a certain kind of illumination in which a camera can be used for image joining and the other can be used for reduction of noise. However, in the example of speech recognition, system gradually monitors and adjusts itself to the variation introduced by the speaker and its allied instruments such as filter and amplifier.

1.1.8.8 Learning in Uncertainty

In a multiagent system, each of the agents is an independent body, which works either independently or under the coordination of others and performs an assignment as a part of the whole task. Each of these independent bodies or agents (having limited resources such as energy, utility, capability, sensor range, speed of movement, decision accuracy, and communication range) are susceptible to uncertain conditions such as failure, energy-exhaustion, traps, deadlocks, and separated from warm (force noncooperation and fully explorative mode till it gets back to the swarm). In these conditions, the original system must accomplish the task without the lost agent(s) and must detect and overcome the incapable or unavailable conditions of uncertain agents through reassignment, rescheduling, and self-reconfiguration of the tasks and according to the requirement of the situations.

1.1.8.9 Knowledge Acquisition in Graph-Based Problems Knowledge

Acquisition is the mode of efficient determination of some solutions or final state or target through the application of some algorithmic steps (deterministic or randomized) and will be called efficient if it takes minimum state transitions for determination of this solution. However, there is always some restriction, which may be actually difficult to represent, and we can find out a certain best portion (set as our goal); this can represent an optimized solution for a certain problem, irrespective of what the constraint is. This is what we need to search out from a graphical representation, a certain combination of nodes and edges, which will hold some intense sequential meaning as well some analytic properties that will elect it as the best possible solution. Similar to the traveling salesman problem (TSP), the most primary requirement is a reduction of the traveled distance. But say there is some cost of traveling associated with each link and is not related to distance factor, and there is an imposition that the salesman can at most spend a certain amount of money for travel [13]. The problem situation changes, and there occurs some more requirement than just the least distance of travel. Let us introduce another constraint that the salesman must travel within a certain time limit (considering that each link has some time for travel parameter and depends on the speed of the transportation used), spend at most a certain amount, and at the same time cover the least distance. So, gradually it is found that the knowledge is getting spread and acquisition of all the knowledge is more important than just temporary relying on local heuristic decisions and greedy approaches.

Further, in this chapter, we have provided a brief overview of varieties of many combinatorial optimization problems and also some graph-based optimization problems. In reality, representation of both types is identical regarding nodes and edges; and hence, later the name can be used interchangeably, indicating all the problems whose representations are in graphs. However, as the graph-based problems are more generalized problem representations, they are dealt and kept separately. The graph-based problems are sometimes very mathematical and may not have any application-based representation in reality and are only be of interest for the mathematicians. However, considering them as a problem or more interestingly as a riddle, they are highly interesting and challenging to solve them efficiently in polynomial time.

1.2 Combinatorial Optimization Problems

Combinatorial optimization problems are a bunch of NP-Hard problems, and each of these problems can have thousands of combinations of its solutions (and hence, perhaps the name) but the best one is that optimized combination, which produces minimum [maximum] scaling on an evaluation [fitness] function or criteria for minimization maximization) problems. With respect to time, combinatorial optimization problems require the nonpolynomial function to determine computation time, which gradually increases nonlinearly with the number of iterations. There are several combinatorial optimization problems, which has been the field of study of many researchers across the globe. The specialty of these problems is that these are much generalized representations of many problems, which one encounter in our day-to-day daily life and unknowingly one compromise with the best optimal solution. Many times due to low dimensionality of the problems, a person can easily seek out the solution manually, but as the problem increases, even computational program scan be time-consuming and memory hungry. With the increase in the number of events, the expenditure may grow unbound, and this is where several computer scientists and management people have tried to get over through quickly optimized solution seeking. The traditional approaches such as branch and bound, greedy methods, and dynamic programming are time-consuming and nonoppurtunistic. With the advent of the bio-inspired and swarm-based computation, these combinatorial optimization problems are tackled with ease and literature has shown their success in achieving near optimal solutions much better than the deterministic ones. The bio-inspired computation algorithms are mainly randomized by coordinated multiagent-based algorithms in which each agent works in cooperation and through exploitation and exploration they can achieve the optimal goal. The ratio of opportunity between exploitation and exploration is decided or rather depends on the user, the number of agents employed, and on the size of the environment and dynamics of the parameters of the environment, and is finally decided through experimentation. Now let us consider the various combinatorial optimization problems one by one.

1.2.1 Traveling Salesman Problem

The TSP is the most generalized combinatorial optimization problem and many others are constrained variants of it. The problem arises from the profit optimization issue of a salesman where he has to cover several cities (places) to accomplish some tasks without revisiting any city and at the same time minimizing the cost of its travel. The following is the mathematical representation of the TSP in Equation 1.1 representing the connection between cities where n is the number of cities and w_{ij}^k is the weight matrix of the edges (from node i to node j and has k number of parameters with $k \in Z$ and $k \geq 1$). The optimized value would be the least value of the summation of the weight of the edges of the graph, which connects all the cities and is actually the course or route followed by the salesman to optimize the distance of travel or the cost involved.

$$\text{Profit} = \text{MIN}\left(\sum_{i,j \in S} w_{ij}^k\right) \text{with } i \neq j \tag{1.1}$$

The TSP involves constraints like along with the distance cost involved must also be least, he must also cover all the cities in less than a certain time such as a time span and each edge is then involved with a time. The equation of such a situation goes like this:

$$\text{Profit} = \text{MIN}\left(\sum_{i,j \in S} w_{ij}^k\right) \text{for} \sum_{i,j \in S}^{i,j \in S} t_{ij} \leq \text{Threshold with } i \neq j \tag{1.2}$$

where t_{ij} is another parameter involved. In this way, several constraints can be included to make it a multiobjective optimization.

1.2.2 Assignment Problem

Assignment problem is the way of assigning p tasks to p executors, and a cost is associated with each pair $C(n,m)$. It can be mathematically represented as

$$\text{Cost} = \text{MIN}\left(\sum_{i=1}^{i=p} C(i,j)\right)$$

where

$$j \in [1, 2, \dots\dots, p] \tag{1.3}$$

and does not repeat.

The cost of executing the task n on m is $C(n,m)$ where $\forall_p$ we can have $n = m$ or $n \neq m$. The optimization can be achieved with respect to minimization of cost of the overall system.

It is one of the fundamental assignment problems without constraint, and if the perfect assignment is made, then we will find that it will automatically reduce the make-span in which it is considered that one executor is assigned only one task. However, if the constraints are imposed then we can have a scenario in which each of the tasks can be fitted to a limited number of executors, and thus we have limited choice for a certain task. Mathematically, we have

$$\text{Cost} = \text{MIN} \left(\sum_{i=1}^{i=p} C_{(i,j)} \right) \tag{1.4}$$

where

$$C_{(i,j)} \exists j \in [1, 2, \dots\dots, p] $$

and does not repeat.

A well-known representation of assignment problem is the bipartite graph in which each edge is connected to one element or node from each set.

1.2.3 Quadratic Assignment Problem

Quadratic assignment problem (QAP) [14] is another assignment problem in which apart from scheduling there involves a cost between any two schedulers, and hence the scheduling must not only be efficient regarding the assignment but also reduce the cost associated with the flow between the scheduler. Mathematically, QAP can be expressed as

$$\text{Cost} = \text{MIN} \left(\sum_{i,j \in S} w_{ij} * d(i,j) \right) \tag{1.5}$$

where $d(i,j)$ is the Euclidean distance between point i and j and w_{ij} is the weight associated with them.

Let us consider that there are N locations and N facilities and each of the location is connected with each of the others and each facility is related to one flow with each other's facilities, the task of QAP is to minimize the sum of the distances multiplied by the flows, which correspond to the facilities placed at those locations.

Say in the hospital there are N buildings and N departments, and there is a flow of people associated between any two departments, then how should the departments be allocated at the buildings so that the maximum people have to cover the minimum distance to reach their destination? Another example can be very-large-scale-integration (VLSI) design and placement of integrated circuits (ICs) in the breadboard, and the challenge is to reduce energy, heat dissipation, and noise that arise from the interconnection in a circuit. This will happen when the devices, which will be mostly used, are placed nearer than the ones, which are placed apart. This is also one type of the QAP. Malls and big markets and their selling strategies and influence exertion introduce QAP such as strategies. They take the statistics of the goods that are sold relatively together. Then they try to place them so apart that when customers, getting one, search for the other, they cover a large part of the store before getting to the other and in the mean time the customers can be influenced to buy something else. Another strategy can be placing two things, sold together, so close that the person buying one can immediately buy the other.

1.2.4 Quadratic Bottleneck Assignment Problem

Quadratic bottleneck assignment problem (QBAP) is another variant of assignment problem. Mathematically, QAP can be expressed as

$$\text{Cost} = \text{MIN}\left(\text{MAX}\left(w_{ij} * d(i,j)\right)\right) \qquad (1.6)$$

where $d(i,j)$ is the Euclidean distance between points i and j and w_{ij} is the weight associated with it. This problem can be visualized by the following a real-life example. Consider that there are N facilities and N locations and the task is to assign them. For each pair of locations, a distance is specified and for each pair of facilities, a weight or flow is specified as the previous problem (e.g., the amount of supplies transported between the two facilities). However, the difference lies in their evaluation and agenda. Here, the problem is to assign all facilities in such locations that there is minimization of the maximum of the distances multiplied by the corresponding flows. Contrary to the task of QAP to minimize the maximum of the distances multiplied by the flows, which correspond to the facilities placed at that locations, in this QBAP, the objective is concerned with each pair of facilities and its corresponding locations and the job is to optimize in such a way that it will minimize all pairs and the maximum of them will also decrease.

1.2.5 0/1 Knapsack Problem

Knapsack problem [15] deals with the optimization of profit to maximum for packing of knapsack with the bounded weight. There are several goods, and each of them is associated with a weight and profit. Each of the sacks is of limited weight. So, the question is—what must be accompanied or packed inside the sack so that the summation of profit is maximized? In addition, either the whole of the item is included in all the sacks or not taken in any of them. There is no intermediate or partial weight that is considered. This makes the problem limited, and convergence can be easily reached having a definite optimum value. This type of knapsack problem is known as 0/1 knapsack problem and is a particular case of the generalized knapsack problem. Mathematically, it can be expressed as

$$\text{Profit} = \text{MAX}\left(\sum_{i=1}^{i=N} x_i p_i\right) \quad \text{for} \sum_{i=1}^{i=N} x_i w_i < W \qquad (1.7)$$

where $x_i \in 0,1 \ \forall_j \in 1,2,......,N$ and p_i is the profit and w_i is the weight associated with ith element. So, the principle is to maximize the profit with bounded weight carrying. N is the number of objects and W is the capacity of the sack. This is valid for one bag. For *multidimensional knapsack problem* the same equation will be applied for each sack and in that case there will be corresponding W, $\{p\}_N$ and $\{w\}_N$ for each sack. Mathematically, we can write

$$\text{Profit} = \text{MAX}\left(\sum_{j=1}^{j=M}\sum_{i=1}^{i=N} x_i p_{ji}\right) \quad \text{for all } j \text{ with } \sum_{i=1}^{i=N} x_i w_{ji} < W_j \tag{1.8}$$

where $j = 1, 2, 3, \ldots, M$ where M is the number of bags.

This kind of problem is relevant to profit-making organizations where they try to maximize the profit with the minimum of work being delivered or performed. Another good example can be for a thief who has visited a house and must carry away only those things whose resale value will fetch him maximum money rather than taking away heavy goods of less resale value. Similarly, for a businessman he would only carry the rare things, which are lighter and have the good profit margin to make money in one go. The problem seems to be easy, but in reality, if the number of such goods is considerably more with many bags of unequal weight and all the goods have a different profit margin, then it will be very difficult for any person to make out what to carry and how much to carry. This makes 0/1 knapsack problem a combinatorial optimization problem.

1.2.6 Bounded Knapsack Problem

Bounded knapsack problem is another kind of knapsack problem in which the mathematical equations are represented as follows:

$$\text{Profit} = \text{MAX}\left(\sum_{i=1}^{N} p_i x_i\right) \tag{1.9}$$

Such that

$$\sum_{i=1}^{N} w_i x_i < W \tag{1.10}$$

where x_i is bounded by $0 \le x_i \le b_i$ and $i \in 1, 2, 3, \ldots, N$. This is unlike 0/1 KSP, here x_i can take positive integer values other than 0 and 1. In that case, both the weight and profit will be multiplied by the factor x_i.

1.2.7 Unbounded Knapsack Problem

Unbounded knapsack problem is a more generalized knapsack problem in which there is no upper bound on x_i, the following mathematical equation represent this:

$$\text{Profit} = \text{MAX}\left(\sum_{i=1}^{N} p_i x_i\right) \tag{1.11}$$

Such that

$$\sum_{i=1}^{N} w_i x_i < W \tag{1.12}$$

where $x_i \geq 0$ for all integral value of i that is $i \in Z$.

1.2.8 Multichoice Multidimensional Knapsack Problem

Multichoice multidimensional knapsack problem is a constrained knapsack problem in which the items are grouped into classes, and there is a bound that exactly one item must be selected from each class. Mathematically, we can represent multichoice multidimensional knapsack problem as

$$\text{Profit} = \text{MAX}\left(\sum_{j=1}^{j=K} \sum_{i \in N_j} x_{ji} p_{ji} \right) \tag{1.13}$$

for

$$\sum_{j=1}^{j=K} \sum_{i \in N_j} x_{ji} w_{ji} < W_j \tag{1.14}$$

where $x_{ji} \in \{0, 1\}$ and K is the number of classes and N_j are the classes and $N_j = \{1, 2, 3, \ldots\ldots, K\}$.

1.2.9 Multidemand Multidimensional Knapsack Problem

For multidemand multidimensional knapsack problem, we have the following representations for the problem and its deciding factor:

$$\text{Profit} = \text{MAX}\left(\sum_{i=1}^{N} p_i x_i \right) \tag{1.15}$$

such that

$$\sum_{i=1}^{g} w_{ji} x_i \leq W_j \quad \text{for } j = \{1, 2, \ldots\ldots, q\} \text{ and } \sum_{i=g}^{N} w_{ji} x_i \geq W_j \quad \text{for } j = \{q+1, 2, \ldots\ldots, m\}$$

where $x_i \in \{0, 1\}$ and $i = \{1, 2, 3, \ldots, N\}$ and N is the number of objects, M is the number of knapsacks and q is any intermediate value between 1 and M.

1.2.10 Quadratic Knapsack Problem

In quadratic knapsack problem, the profit is calculated twice, once for p_{ij} and another for p_{ji} and if symmetric then we have $p_{ij} = p_{ji}$. Here, the profit matrix is $n \times n$, and hence named quadratic. If item i and j are selected then profit is $p_{ij} + p_{ji}$

$$\text{Profit} = \text{MAX}\left(\sum_{i=1}^{N} \sum_{j=1}^{N} q_{ij} x_i x_j \right) \tag{1.16}$$

with

$$\sum_{i=1}^{N} w_i x_i \le W \wedge x \in \{0,1\} \tag{1.17}$$

1.2.11 Sharing Knapsack Problem

In sharing knapsack problem, the N objects are divided into m classes and J_j is any jth class where $j_j \subseteq \{1, 2, \ldots, N\}$ and each object i is associated with profit p_i and weight w_i

$$\text{Profit} = \text{MAX}\left(\underset{1 \le j \le m}{\text{MIN}}\left(\sum_{i \in J_j} p_i x_i\right)_j\right) \tag{1.18}$$

Subjected to

$$\sum_{i=1}^{N} w_i x_i \le W \tag{1.19}$$

where $x_i \in \{0,1\}$ for $i = \{1, 2, \ldots, N\}$. In sharing knapsack problem, the maximization occurs for the least among all the classes so that there is adequate participation from each class. Say in an organization some prize money is to be distributed among the employees and the employees are grouped according to position. Each employee's prize, if he gets, is proportionate to his group. We should try that the total amount of prize money given to each group is almost the same or if not the same then at least try to maximize the minimum for any group. This is the philosophy for sharing knapsack problem.

1.2.12 Corporate Structuring

Corporate structuring is another NP-Hard optimization problem that decides on which are the countries the company must target to extend their centers. There are a few parameters associated with each country such as taxes and expected profit. In addition, there are constraints that the country with net profit greater than zero must be included, but countries with zero profit may or may not be included. Mathematically,

$$\text{Profit} = \text{TaxCode} + \text{ForeignIncomeTaxRate} + \text{DomesticIncomeTaxRate} + \text{Profit}$$
$$+ \text{WithHoldingTaxRate} \ge 0$$

where

$$\text{WithHoldingTaxRate} = \sum_{j=1, j \ne i}^{j=N} \text{whtr}_{ij} \quad \text{for } j \in S \wedge S \subseteq C'$$

and whtr_{ij} is the WithHoldingTaxRate for the country i with the country j and $j \in \{1, 2, \ldots, n\}$ where n is the number of countries other than i.

1.2.13 Sequential Ordering Problem

Sequential ordering problem (SOP) [16] tries to make a sequence out of a set of events or nodes or objects so that the precedence of any event is not violated and at the same time the criteria of maximization of profit (minimization of cost) are maintained. The following example will make the conception clear and it will provide a visualization of how the problem is represented and how to handle the datasets of SOP. Consider the following matrix that provides the information for evaluation of solution for an ordered sequence.

Each of the positive values represents the w_{ij} that is the weight of the edge from node i to node j and the rest of the individual elements having values $w_{ij} = -1$ indicates that the node i cannot occur before node j. In other words, it can be said that the event i cannot occur before event j. Fitness evaluation procedure is same as TSP but the sequence needs to be evaluated and validated, and only the acceptable sequence will be considered for fitness evaluation.

This kind of problem can be regarded as a constrained version of the TSP in which all the events need to be incorporated and at the same time, the precedence of the events is respected. For example, in a plant (or during software development) some processes need to be completed before the other events. Likewise, say for a logistic company, some pickups are required, and many deliveries are there, so it is better to visit the pickups before visiting the places where those pickups need to be delivered. In such cases, the cost of travel must be minimized to enhance profit.

1.2.14 Vehicle Routing Problem

The objective of vehicle routing problem (VRP) is to cover several places within a city for task accomplishment with a certain number of vehicles such that the cost associated with the coverage is at the minimum. The example can be a logistic company that wants to deliver to the customers, and they have some vehicles. They are going to figure out which vehicle is going to visit where so that the travel cost is at the minimum. Mathematically, VRP [17] can be expressed as

$$\text{Cost} = \text{MIN}\left(\sum_{i \in V} \left(\sum_{j,k \in C_i} d(j,k) \right) \right) \text{with } j \neq k \qquad (1.20)$$

where V is the number of vehicles, $C_i \subseteq C$ with C_i are the places visited by the ith vehicle. There are several examples of VRP such as logistic delivery planning, dropping of employee by the cars in an organization or school, and power/water supply distribution by multiple centers.

1.2.15 Constrained Vehicle Routing Problem

Constrained vehicle routing problem (CVRP) is another version of the VRP having the constraints that must be satisfied, and, at the same time, the overall cost must be minimized. The minimization of cost with an unsatisfied constraint condition is not acceptable. There are two kinds of CVRP, one for the load and another for the precedence maintainability. In the case of load constraint, each vehicle has a load carrying limit and each of the places is associated with one load to deliver. So, the cost minimization is considered if and only if each of the used vehicles does not exceed the load limit that it can carry. For the maintainable precedence graph, there is a requirement of visiting of one place before another by the same vehicle. This scenario is common for logistic companies, which need to pickup goods from one place and deliver it to another place. However, the second constraint problem is highly complex regarding both fitness evaluation and deciding solutions. Mathematically, CVRP has similar equations such as VRP with some extra conditional equations.

$$\text{Cost} = \text{MIN}\left[\sum_{i \in V}\left(\sum_{i,k \in C_i} d(j,k)\right)\right] \quad \text{for } j \neq k \text{ subjected to } \sum_{l \in C_i} w(l) \leq W_i \tag{1.21}$$

where, V is the number of vehicles, $C_i \subseteq C$ with C_i are the places visited by the ith vehicle. $w(l)$ are the weights associated with C_i for the vehicle i and the summation of the weight must be equal or less than W_i that is the capacity of the vehicle i.

There are also similar examples of CVRP similar to that of VRP such as logistic delivery planning, dropping of employee by the cars in an organization where the vehicle is a car and has limited number of seats, and power/water supply distribution by multiple centers where each power center can serve limited clients.

1.2.16 Fixed Charge Transportation Problem

Fixed charge transportation problem consists of one fixed cost and another variable cost proportional to the weight carried. The route foundation must minimize the total of the variable and the fixed cost, whereas the demand and supply load are satisfied for each source–destination pair. Mathematically, we can represent the fixed charge transportation problem as

$$\text{Cost} = \text{MIN}\left(\sum_{i=1}^{S}\sum_{j=1}^{D}\right)\left(v_{ij}x_{ij} + f_{ij}y\left(x_{ij}\right)\right) \tag{1.22}$$

Such that we have

$$\sum_{j=1}^{D} x_{ij} = s_i \text{ and } \sum_{i=1}^{S} x_{ij} = d_j \tag{1.23}$$

where $i = \{1,2,\ldots\ldots,S\}$ and $j = \{1,2,\ldots\ldots,D\}$, respectively, and

$$y(x_{ij}) \in [0,1] \text{ and } x_{ij} \geq 0, \forall_i \in S, \forall_j \in D \tag{1.24}$$

where, we have S is the number of supply points, D is the number of demand points, S_i is supplied for origin i, d_j is the demand of destination j, f_{ij} is the fixed cost associated with each unit transferred from node i to node j, v_{ij} is the variable cost incurred to switch unit from node i to node j, x_{ij} is the amount of load transferred from node i to node j, $y(x_{ij}[0,1])$ is the factor to include the fixed cost if load is present that is if $x_{ij} > 0$.

1.2.17 Job Scheduling

Job scheduling problems [18,19] is the process of optimization of the make-span (total execution time of the schedule) of several jobs of varying length through two or more processing or execution units. Simultaneous processing will bring down the parallel execution time but the overall make span is required to be reduced, and further optimization through proper strategy will reduce the overall completion time. Mathematically, job scheduling [20] can be expressed as

$$
\text{MakeSpan} = \text{MIN}\left(\text{MAX}\left[\left\{ \sum_{i\in S_1} t_{J_i} \right\} \cup \left\{ \sum_{i\in S_2} t_{J_i} \right\} \cup \ldots\ldots\ldots \cup \left\{ \sum_{i\in S_n} t_{J_i} \right\} \right] \right) \qquad (1.25)
$$

where $\{J_1, J_2, \ldots\ldots\ldots, J_N\}$ is the set of N jobs to be assigned for n execution units and nth execution unit is assigned S_n jobs, and t_{J_N} is the execution time taken by the J_i job.

1.2.18 One-Dimensional Bin Packing Problem

One-dimensional bin packing problem is another NP-Hard problem described by the optimization requirements for minimization of number of bins to accommodate several objects of definite volume. The volume of the bin cannot be exceeded and at the same time for each bin the residual space must be minimized, and that is how the number of bags can be minimized.

$$
\text{Bin} = \text{MIN}n = \text{Count}\left[\sum_{i\in S_1} v_i, \sum_{i\in S_2} v_i, \ldots\ldots\ldots\ldots, \sum_{i\in S_n} v_i \right] \qquad (1.26)
$$

where n is the minimum bags required v_i is the volume of the ith item or item a_i, S_m is the mth set consisting of certain objects and $S \subseteq a_1, a_2, \ldots\ldots, a_N$ where N is the number of objects and each a_k is having corresponding volume v_k. Also we have the condition

$$
\sum_{i\in S_k} v_i \leq V \qquad (1.27)
$$

where V is the capacity of each bin and k is the kth bin involved.

An example of bin packing problem can be the loading of vehicles with goods and the requirements of a minimum number of vehicles. Another can be container filling with items in the kitchen where the requirement is to have a minimum number of containers.

1.2.19 Two-Dimensional Bin Packing Problem

Two-Dimensional bin packing problem is an extension to the one-dimensional version of the bin packing problem. Here, each of the items has both length and width (but of equal height) and they are to be packed in bins, and the optimization must decrease the number of bins. The two-dimensional bin packing problem can also be regarded as a cutting stock problem; the difference is that in the former the objects are placed on a stock and in the latter the objects are cut from the stock.

$$
\text{Bin} = \text{MIN}\left(\text{Count}\left(\sum p_i \right) \right) \qquad (1.28)
$$

where $p_i = s_i \subseteq S$ and $p_i \cap p_j = \varnothing$ where $i \neq j$, and it means that from p_i we can create all s_i jobs with minimum wastage, where $S \in \{x_1, x_2, \ldots\ldots\ldots, x_n\}$ and represents all the n jobs and n is the number of jobs.

1.3 Graph-Based Problems

Graph-based problems are the basic problems whose subparts are combinatorial optimization problems discussed earlier. The only difference between the two is that combinatorial optimization problems are specific examples and are easy to understand, but the graph problems are abstract and involve mathematical challenges and have various applications in real world. The graph theory, which monitors and governs the graph-based problems, mainly deals with exploration, research, and identification of the various properties, characteristics, and problems of the graphs and algorithms that are related to solving them. In this chapter, as a part of the introduction to the various graph-based problems that can be solved and optimized with the nature-inspired algorithms, we have introduced only some specific cases that are dealt with. The field of graph theory is vast and beyond the scope of this book.

1.3.1 Graph Coloring Problem

The graph coloring problem [21] is one kind of optimization problem which finds the least number of colors required to color each node of the graph such that none of the adjacent nodes get the same color. The problem is NP-Hard and opportunistic, and there can be several valid solutions, though very less number of optimized solutions. The solution can be represented as a string or chromosome, and the evaluation function must ensure that the adjacent nodes are not of the same color. However, there is no mathematical representation or equation for the graph coloring problem.

1.3.2 Path Planning Problem

Path planning problem deals with establishing a sequence of nodes between any two nodes of the graph that must be optimized for some parameters or the number of nodes or any other special constraints, which makes up maximum profit out of the existing situation. However, general path planning problem deals with finding the shortest path between the source and the destination. However, if each node is regarded as an event, the optimized sequence of events will create least time span between the two events, which can be helpful in process design in the manufacturing unit of a factory or may be in time calculation for software development. Path planning problem can also be referred to the path planning [22] for robots in some continuous domain search space in which each point is represented as a node or a grid and is simultaneously connected to eight other nodes or grids. Such movement-based search and establishment of path optimized with distance is also a graph-based problem. Mathematically, path planning problem can be represented as

$$\text{Cost} = \text{MIN}\left(\sum_{i=1}^{i=n-1} W(a_i, a_{i+1}) \right) \tag{1.29}$$

where n is the number of nodes in the path, whereas $W(j_1, j_2)$ denotes the distance weight between the node. a_i and a_{i+1}, If considered very accurately resource constraint shortest path problem (RCSP), SOP, and so on, are the one kind of constrained path planning problems.

1.3.3 Resource Constraint Shortest Path Problem

RCSP is a constrained graph-based path planning problem, which operates or validates under the satisfaction of some mathematical inequality. Mathematically, we can also represent RCSP as the Equation 1.29, but until the following condition is satisfied:

$$\sum_{i=1}^{i=n-1} C(a_i, a_{i+1}) < \text{Threshold} \tag{1.30}$$

where $C(a_i, a_{i+1})$ is the parameter weight for constraint C involved on the edge between the nodes a_i and a_{i+1} and there is some threshold associated with that parameter. The less than sign in Equation 1.30 can also be greater than the sign that depends on the problem model. Rest of the variables are same as in Equation 1.27. There can be various such conditions depending on the system model and parameters involved. This makes the problem a multiobjective optimization. There can be numerous constrainted versions of the graph-based path planning problem RCSP is just a special case.

1.4 Aim of This Book

The aim of this book is to provide budding researchers, a better view of nature-inspired algorithms, along with their variants, their analysis, and the details how they can move forward to seek a solution to a variety of problems with the help of these algorithms. The advantages of the variants are given from a bird's eye view. There is a wide range of work on discrete optimization, as the next-generation problems will deal with a graph such as discrete problems. However, for completeness, both continuous domain problems, as well as discrete domain problems are discussed to show clearly that the algorithms have evolved from the continuous domain to the discrete domain and, are being applied for various other real-time domain problems. This book can be used by both the undergraduate students who have started digging into this field of bio-inspired algorithms and its applications as well as for graduate students as a reference. However, this book is more intended toward the discrete problems as majority specialization similar to that of the path formation in graphs, combinatorial optimization problems, along with other graph-based problems. The introduction of model-based examples will provide a detailed view of how problems are handled and modeled into some favorable mathematical representation for a quick convergence of the solution. Hybrid structures have been very successful in many aspects and have also taken up to analyze how the best of the algorithms are mixed for specific problems.

1.5 Summary

In this chapter, we have mainly focused on the introduction of the different subfields of the artificial intelligence and its allied branches. We have also introduced the various discrete problems such as graph problems and combinatorial optimization problems, most of which are dealt with in the subsequent chapters. In addition to it, some real-life examples and their representation as discrete problems are also discussed for the sake of generality.

Solved Examples

1. Distinguish between an informed search and an uninformed search

   Uninformed search

   This determines the graph and the goal but not which path to select from the frontier. This remains the job of a search strategy. A search strategy then specifies which paths are selected from the frontier. Different strategies are obtained by modifying how the selection of paths in the frontier is implemented.

   a. Depth-first search

   b. Breadth-first search

   c. Lowest cost first

   Informed search

   It is easy to see that an uninformed search has a tendency to pursue options that lead away from the goal as easily as it pursues options that lead toward the goal. For any but the smallest problems, this leads to searches that take unacceptable amounts of time and/or space.

   As an improvement over this, an informed search tries to reduce the amount of search that must be done by making intelligent choices for the nodes that are selected for expansion. This implies the existence of some way of evaluating the likelihood (using a heuristic function) that a given node is on the solution path.

   a. Best Fit search

   b. A* search

2. Discuss the rationale behind the principle of least commitment

   The principle of least commitment is the idea of never making a choice/decision unless required at the moment. The idea is to avoid taking decisions that may be reverted at a later stage. Thus, it goes a long way in avoiding wastage of computing facilities and at the same time ensures that the algorithm runs smoothly. This principle is used in many artificial intelligence algorithms; for example, the partial order planner algorithm uses this principle to delay the assignment of order in which the actions need to be taken.

3. Discuss the multiagent framework in artificial intelligence

   A multiagent framework is characterized by the following:

   - The agents can act autonomously, each having its own information about the world/environment and the other agents.

   - The outcome depends on the actions of all the agents. A mechanism function defines how the actions of individual agents lead to a collective outcome.

   - Each agent can have its own utility that depends on the outcome.

Each agent decides as what to do based on its own utilities, but it must also interact with other agents. An agent acts strategically when it decides what to do based on its goals or utilities.

There are two extremes in the study of multiagent systems:

- *Fully cooperative*: Agents share the same utility function.
- *Fully competitive*: When one agent can win only when another loses. These are often called zero-sum games in which the utility can be expressed in a form such that the sum of the utilities for the agents is zero for every outcome.

4. Comment on the advantages and disadvantages of group decision-making

Advantages

a. *Synergistic approach* to performing a task: Synergy is the fundamental notion that the *whole is greater than the sum of its parts*. When a group of robots make a decision collectively, its judgment can be keener and more effective than that of any of its members.

b. *Information sharing*: Sharing information can increase the understanding, clarify issues/misinterpreted facts, and facilitate an action that is in coherence with the collective decision.

Disadvantages

a. *Diffusion of responsibility*: leading to lack of accountability for the nonacceptable outcomes.

b. *Lower efficiency*: Group decision-making process can take additional time because there is an increased overhead of participation, discussion, and coordination among the agents.

c. The group thinks the facts known to individuals can conflict and there is no guarantee that the collective decision is the best decision possible.

5. Discuss the components of regression planning

Regression planning is based on graph search and traversals in order to find the optimal paths. Its various components include the following:

- The nodes are goals that must be achieved. A goal is a set of assignments to (some of) the features.
- The arcs/edges correspond to actions. In particular, an arc from node g to g', labeled with the action *act*, means that the *act* is the last action that is carried out before the goal g is achieved, and the node g' is the goal that must be true immediately before the *act* so that g is true immediately after the *act*.
- The start node is the goal to be achieved.
- The goal condition for the search, *goal(g)*, is true if all of the elements of g are true of the initial state.

Given a node that represents goal g, a neighbor of g exists for every action *act* such that

- *Act* is possible: It is possible for the *act* to be carried out and for g to be true immediately after the *act*.
- *Act* is useful: *Act* achieves part of g.

6. Discuss the concept of Partial-order planning

   Partial-order planning is an approach to automated planning that delays the acts of action ordering as much as possible. Its approach is in contrast with the total-order planning that produces an exact ordering of actions. Partial-order plan specifies all actions that need to be taken, but specifies an ordering of the actions only where necessary.

   A partial-order plan consists of four components:

   - A set of actions (also known as operators).
   - A partial order for the actions. It specifies the conditions about the order of some actions.
   - A set of causal links/bindings. It specifies which actions meet which preconditions of other actions.
   - A set of open preconditions that specify which preconditions are not fulfilled by any action in the partial-order plan.

7. Discuss the role of Heuristic functions in artificial intelligence

   A heuristic function is a function that maps from problem state description to measure desirability; the mapping being represented as number weights. These functions need to be chosen carefully as well-designed heuristic functions can provide a fairly good estimate of whether a path is good or not. Thus, a heuristic function guides the search process in the most profitable directions, by suggesting which path to follow first when more than one path is available. However, in some problems, the cost of computing the value of a heuristic function can become more than the effort saved in the search process. In these situations, a trade-off becomes a necessity.

8. Discuss the "exploration problem" in the context of the multi-robot system

   This problem deals with the use of a robot to maximize the knowledge that it can gather over an area. The exploration problem typically arises in mapping and search and rescue situations, where an environment might be dangerous for the proper functioning of the agent. This is also one of the reasons why a great deal of research goes into multirobot systems (MRS). These systems have distinct advantages as compared to its peers:

   a. Reduced exploration time, multiple robots can explore the area faster than one robot.
   b. Improved efficiency of mapping.
   c. Success of mission ensured in all conditions.
   d. Forward compatible technology.

Exercises

1. Discuss the motivation behind using bio-inspired algorithms.
2. Discuss occupancy factor and its role in soft computing algorithms.

3. In what circumstances, can artificial intelligence outnumber human intelligence?

4. In what class of problems does randomness appear to be more useful than more deterministic approaches?

5. How is an intelligent system different from an expert system?

6. Discuss the rationale behind inference systems.

References

1. J. I. Fister, X. S. Yang, I. Fister, J. Brest and D. Fister, A brief review of nature-inspired algorithms for optimization, *Elektrotehniski vestnik*, 2013, 80, 116–122.

2. M. Kubat and R. C. Holte, Machine learning for the detection of oil spills in satellite radar images, *Machine Learning*, 1998, 30, 195–215.

3. C. Andrieu, N. D. Freitasa, A. Douchet and M. I. Jordan, An introduction to MCMC for machine learning, *Machine Learning*, 2003, 50, 5–43.

4. C. M. Bishop, *Pattern Matching and Machine Learning*, Springer, New York, 2006.

5. E.-G. Talbi, A taxonomy of hybrid metaheuristics, *Journal of Heuristics*, 2002, 8, 541–564.

6. J. S. R. Jang, C. T. Sun and E. Mizutani, Neuro-fuzzy and soft computing—A computational approach to learning and machine intelligence, *IEEE Transactions on Automatic Control*, 1997, 42, 1482–1484.

7. M. Stefik, Retrospective on "The organization of expert systems, a tutorial", *Artificial Intelligence*, 1993, 59, 221–224.

8. D. H. Wolpert and W. G. Macready, No free lunch theorems for optimization, *IEEE Transactions on Evolutionary Computation*, 1997, 1(1), 67–82. doi:10.1109/4235.585893.

9. G. Nemhauser and L. A. Wolsey, *Integer and Combinatorial Optimization*, Wiley, New York, 1988.

10. W. van der Hoek and M. Wooldridge, Multi-agent systems, In B. Porter, V. Lifschitz and F. V. Harmelen (Eds.), *Handbook of Knowledge Representation*. Elsevier, Amsterdam, the Netherlands, 2007.

11. I. A. Taharwa, A. Sheta and M. A. Weshah, A mobile robot path planning using genetic algorithm in static environment, *Journal of Computer Science*, 2008, 4(4), 341–344.

12. A. Stentz, Optimal and efficient path planning for partially-known environments, *Proceedings IEEE International Conference on Robotics and Automation*, San Diego, CA, May 8–13, 1994.

13. R. Bellman, Dynamic programming treatment of the travelling salesman problem, *Journal of the Association for Computing Machinery*, 9, 61–63.

14. E. L. Lawyer, The quadratic assignment problem, *Institute for Operations Research and the Management Sciences*, 1963, 9(4), 586–599.

15. C. Chekuri and S. Khanna, A polynomial time approximation scheme for the multiple knapsack problem, *SIAM Journal on Computing*, 2006, 35, 713–728.

16. L. M. Gambardella and M. Dorigo, An ant colony system hybridized with a new local search for the sequential ordering problem, *INFORMS Journal on Computing*, 2000, 12(3), 237–255.

17. G. Laporte, The vehicle routing problem: An overview of exact and approximate algorithms, *European Journal of Operational Research*, 1992, 59, 345–358.

18. A. S. Manne, On the job scheduling problem, *Institute for Operations Research and the Management Sciences*, 1960, 8(2), 219–223.

19. D. Applegate and W. Cook, A computational study of the job-shop scheduling problem, *ORSA Journal on Computing*, 1991, 3, 149–156.

20. M. D. Amico and M. Trubian, Applying tabu search to the job-scheduling problem, *Annals of Operations Research*, 1993, 41, 231–252.
21. H. S. Wilf, Backtrack: An O(1) expected time algorithm for the graph coloring problem, *Information Processing Letters*, 1984, 18, 119–121.
22. R. Hassin, Approximation schemes for the restricted shortest path problem, *Institute for Operations Research and the Management Sciences*, 1992, 17(1), 36–42.

2

Particle Swarm Optimization

2.1 Introduction

Particle swarm optimization (PSO) is the pioneer of swarm intelligence algorithms, which utilizes collective effort for achieving better convergence for optimization. It was introduced by J. Kennedy and R. Eberhart in the year 1995 [1,2]. The history of the development of PSO is quite interesting as the two founders came from diverse backgrounds and they had been cooperating for sociopsychological simulation purposes and the outcome is one of the finest algorithms for optimization and problem-solving. The PSO algorithm mainly works on the basis of two factors, one is the collective exploration and the second behavior is the social guidance-based influence of the coparticles for decision-making of the other particles.

The particles perform a search based on the evaluation function taking into consideration the topological information (which will ultimately transform into the solution set of the problem) and share their knowledge and success evaluation with the other particles iteratively before they, as a whole or individually, converge to the optimum solution. The social influence of the PSO is being formulated through some mathematical foundations, which constitute and represent the social behavior and the mode of decision-making for the particles for subsequent iterations based on the present situation. The two main equations, which govern the agents of PSO and represent the intelligent collective behavior, are given as follows:

$$x_{i(t)} = x_{i(t-1)} + v_{i(t)} \tag{2.1}$$

We can represent x_i as $\{x_1, x_2, \ldots, x_n\}$ where $x_i \in S$ (Solution set) and n is the dimension of the solution. Here, we have a velocity component as $v_i \forall x_i$ and is interpreted as the variation component for the x_i axis and v_i can be calculated as

$$v_i(t) = wv_i(t-1) + r_1 c_1 \left(x_{ib} - x_i(t-1) \right) + r_2 c_2 \left(x_{gb} - x_i(t-1) \right) \tag{2.2}$$

where "t" represents the current iteration or rather the tth iteration. The two equations are iterative and require some memory as they utilize stored values from the previous iteration. In the equations, we have certain weighted parameters and some random factors, which introduce variations in the mode and magnitude of the search and create a perfect exploration scheme for the environment. Here, we have r_1 and r_2 as random factors uniformly distributed in the range [0,1] in which we can include or exclude the value zero.

However, theoretically it is justified to include or exclude the social influence of the fellow best particles. In addition, c_1 and c_2 are the two acceleration coefficients that scale the influence of the iteration best and the global best positions. Here, w is an inertia weight scaling of the previous time step velocity factor that must be added to get the new position of the particle. This inertia weight w can also be varied, combined with a momentum or even adaptively increased or decreased. Further, the iteration best i_{best} can be replaced by p_{best}, which is the best solution visited by it so far. The difference between i_{best} and p_{best} is that i_{best} is the best solution for the previously completed solution pool (which may or may not be the global best but sometimes might be when a new global solution is found and updated), whereas p_{best} is the best solution visited by that particular particle (a very less chance exists here that p_{best} be the global best as well). In addition, p_{best} will be different for different particles, whereas i_{best} will be same for all the individuals.

At each iteration, each of the PSO particle stores in its memory the best solution visited so far, called p_{best}, and experiences an attraction toward this solution as it traverses through the solution search space. As a result of the particle-to-particle information, the particle stores in its memory the best solution visited by any particle, and experiences an attraction toward this solution, called g_{best}, as well. The first and second factors are called cognitive (p_{best}) and social (g_{best}) components, respectively. After each iteration, the values of p_{best} and g_{best} are updated for each particle; in case a better or more dominating solution (in terms of fitness evaluator) is found.

2.2 Traditional Particle Swarm Optimization Algorithm

PSO algorithm works on the basis of attraction and detraction components that it acquires from the social influence of the global best and the iteration best particles. However, the influences are magnified (positively and negatively) by two factors r and c, which are known as the random factor and acceleration coefficient, respectively. The range or the absolute value of these two factors varies from problem-to-problem and mainly depends on the magnitude and scope of the search space. In addition, the calculated adaptivity of these two factors has been proven to be very efficient for convergence and has also been contributed in the form of variants of the PSO algorithm.

In that case, the value of c changes gradually (gradually increasing, gradually decreasing, intelligently changing based on some calculated factors, etc.) and r being already a randomly generated one, their product can bring about better variations for the individual variables. Some such variants will be described in the subsequent sections. Some of them are more biased and more converging, whereas many of them are explorative themselves. However, a combination of such exploitative and explorative features can bring about the best out of the working principle of the mathematical expressions of the PSO algorithm.

The main advantages of the PSO algorithm are its simple concept and ease with which one can visualize and frame the problem for solution representation, easy implementation, relative robustness to control parameters, their ranges and constraints, and the computational convergence efficiency. Although the PSO algorithm-based approaches have several advantages, they may get trapped in a local minimum for multimodal search spaces and when handling heavily constrained problems due to the bias of the social influence and less individual exploration culminating in the limited local/global searching and decision-making capabilities. The various drawbacks of the PSO algorithm such as local optimal solution trapping leading to premature convergence, insufficient capability to find very nearby extreme points due to lack of iterative variation (i.e., exploitation problem), and lack of efficient mechanism to treat the constraints (i.e., constraint handling problem) can be tackled by additional operations. Many of the exploration and exploitation criteria are handled by introducing new variants, added parameter control, adaptive learning, and influencing techniques. Many of those variants that have been proven to be better regarding efficiency and convergence rates are discussed in the subsequent sections. The whole PSO algorithm, in a nutshell, can be represented in the following algorithmic steps:

Steps for the Traditional Particle Swarm Optimization Algorithm

---

Procedure (PSO) (Problem$_{Dimension}$, agents, Constraints, Search$_{Space}$)
Initialize N Agents with D dimension represented as $\{x_1, x_2, \ldots, x_D\}$
Initialization of Agent Memory, and Initialize it to random value
Evaluate Fitness of Agents
if Best then (N Agents) is.BetterGlobal.Best
Update Global.Best
end if
for Iteration $i = 1$ to Iteration$_{max}$ do
for Agents $j = 1$ to N do
for Agent's Dimension $k = 1$ to D do
Calculate r_1, c_1, r_2, c_2
Calculate $v_{jk}(i) = w * v_{jk}(i-1) + r_1 c_1 (x_{gb}(i-1) - x_{jk}(i-1)) + r_2 c_2 (x_{ib}(i-1) - x_{jk}(i-1))$
Calculate $x_{jk}(i) = x_{jk}(i-1) + v_{jk}(i)$
end for
end for
if Best then (N Agents) is.BetterGlobal.Best
Update Global.Best
end if
end for
end procedure

2.3 Variants of Particle Swarm Optimization Algorithm

Several variants of the PSO algorithm have been proposed so far, using which optimizations have been performed on several types of multidimensional applications [1]. The various improved variants of the traditional PSO algorithm will be discussed in this section, and their features will be illustrated. It will be highlighted how the main variations and introduced operational features have been better for understanding the underlying dynamics of the particles of the algorithm and exactly how the performance of the algorithm has been for various applications.

2.3.1 Sequential Particle Swarm Optimization Algorithm

If we have a varying or moving modal search space in which the global minima is constantly moving, we cannot start the search from the beginning. At the same time, the search process and exploration must continue. Sequential PSO algorithm is one such variant of PSO that has been implemented for tracking objects in a medium (image or video). Traditional PSO algorithm always starts from the initialization as it preserves some global and personal (iteration) best. However, as the search space has changed, global and personal (iteration) best values become vague, and there is a requirement of reinventing the situation.

Sequential PSO algorithm [3,4] is a variant of PSO algorithm mainly introduced to solve a particular application of visual tracking. In visual tracking, the object is in motion, and hence the search environment changes with time. In this kind of random and dynamic problem, the optimization in local optima requires special care and is very difficult to handle with the traditional PSO algorithm. Hence, the modified sequential PSO algorithm is introduced with special operations to take care of the changing environment. The main idea behind the introduction of this variant is to encourage its involvement in more variations, diversities, and learning factors based on the measured experiences for the search space involving the progress of the search taken into consideration. The previously found best individuals on optimization are initially randomly propagated to enhance their diversities, and this is followed by the operations of sequential PSO algorithm with adaptive parameters that are constantly tuned on iteration and coagents. Finally, the stopping criterion is evaluated to check whether the algorithmic iteration convergence criterion is met based on the new operation of an efficient convergence criterion. The three main operational steps of the sequential PSO algorithm framework are as follows:

1. Random propagation
2. Initialization of adaptive PSO
3. Convergence criterion check

The details of the operations will now be elaborated with details and example descriptions.

2.3.1.1 Random Propagation

In the traditional PSO algorithm, the convergence of the previous particles tends to face difficulty when a new search space with different optimization criteria is introduced. The major difficulty lies in the diversity loss of particles due to the previous iteration best and global best particles, which keep on influencing the coagents for movement and decision-making. Hence, a rediversification mechanism is employed in the sequential PSO algorithm to reinitialize virtually the agents for the next image frame. An effective rediversification mechanism needs to keep track of the gained prior knowledge of the motion of the object being tracked.

In the sequential PSO algorithm particle representing the solution set for the problem is randomly propagated following the traditional decisive Gaussian transition model in which its mean is considered as the fitness of the previous individual best particle. The covariance matrix of the generated sequential PSO algorithm particle solutions is determined by the magnitude and the direction of the predicted velocity of the object motion.

In the rediversification strategy for the random propagation of particles, resampling process is not needed because the individual best of the particle set converged at time t provides a compact sample set for the propagation. Although random propagation depending on the predicted velocity is simple, it is quite sufficient because the main purpose is only to initialize some values for commencing a subsequent optimal search.

2.3.1.2 Adaptive Schema for Sequential Particle Swarm Optimization Algorithm

According to the author [4], to counteract the possibility of fostering the danger of swarm explosion and divergence especially in high-dimensional state space, some reasonable mechanism for controlling the acceleration parameters such as φ_1, φ_2, and the maximum velocity $v_{max}(t)$ are being used. The two acceleration parameters φ_1, φ_2 are defined as the following:

$$\varphi_1 = \frac{2 * \text{fit}(\text{Part}_i)}{(\text{fit}(\text{Part}_i) + \text{fit}(gb))} \tag{2.3}$$

$$\varphi_2 = 2 * \frac{\text{fit}(gb)}{\text{fit}(\text{Part}_i + \text{fit}(gb))}$$

where fit() is the fitness function and Part_i is the ith particle, and gb is the global best particle.

The use of these self-tuning parameters in modified sequential PSO algorithm will overcome the drawback of the lack of reasonability for the dynamic environment. In traditional PSO algorithm, the parameters are constant sets and hence unreasonable. This self-adaptive scheme will demonstrate the use of *cognitive* part or *social* part determined by their fitness values. As seen in the applications of object tracking the maximum velocity $v_{max}(t)$ provides a reasonable bound to cover the particle's maximum motion and prevent the particle from arbitrary and random moving schemes. The maximum velocity $v_{max}(t)$ can be set to a predefined constant. However, it is not reasonable when we consider that the object has an arbitrary motion. Therefore, $v_{max}(t)$ is selected based on the predicted velocity $v_{pred}(t)$ of the optimum as the following:

$$v_{max}(t) = \text{Const} * v_{pred}(t) \tag{2.4}$$

where Const is a determined constant.

In the following way, the maximum velocity $v_{max}(t)$ of the particles of the sequential PSO algorithm is heuristically selected by utilizing the motion information of the object in the previous tracking process and thus provides a reasonable limitation to the movement and direction of search for the sequential PSO algorithm particles. Hence, they are made capable of utilizing their acceleration for the dynamically changing environment or search space.

2.3.1.3 Convergence Criterion

The object tracking problem is a special kind of dynamic problem and hence, there is a need to introduce a unique convergence rate for the particle. Expecting that it is not necessary for all the particles to converge to optima and track down the object. In addition, this will help in the better exploration of the search space. The exploration rate must be considerably high, so as to locate the object as quickly as possible and thus produces a better convergence rate. Hence, the convergence criterion is formulated as follows:

$$F(gt) > \text{Th} \tag{2.5}$$

where Th is a predefined threshold, and all the individual best particle with a solution set as $\{x_1, x_2, \ldots, x_D\}$ at time t is in a neighborhood of gb, that is, the global best, or the maximum iteration number is encountered. According to this criterion, the object to be searched can be efficiently identified and the convergent particle with a particle set as $\{x_1, x_2, \ldots, x_D\}$ at time t provides a compact initialization without sample impoverishment for the next optimization process, and the temporal continuity information can be naturally incorporated into the sequential PSO algorithm framework.

The main feature that must be noticed in sequential PSO algorithm is the introduction of an adaptive and explorative feature that will predict the movement of the global best and also involve simultaneous search so that it can keep track of the nearby optima.

2.3.2 Inertia Weight Strategies in Particle Swarm Optimization Algorithm

Inertia weight strategies have targeted the weighted value that gets multiplied as inertia with the present velocity value. Variation inertia weight w in the velocity equation of the traditional PSO algorithm remains static most of the time. If it is very less, exploration is hindered, and if it is large then local search is hampered. Hence, different mathematical expressions have been tried to evolve some changing adaptive features into the overall equation system. This is another way of introducing adaptive and nonuniform variations for the particles of PSO algorithm, [5] has discussed several such schemes and each has its performance and convergence rate, which they can exhibit for different numerical optimization problems through the use of stochastic decision-making.

2.3.2.1 Constant Inertia Weight

The most simplified and nonadaptive mode is the use of a constant value of w_i for any iteration i.

$$w_i = c \tag{2.6}$$

where c is the constant, which needs to be initiated, and the best c can be determined only through experimentation.

2.3.2.2 Random Inertia Weight

$$w_i = a + \frac{\text{rand}()}{b} \tag{2.7}$$

where a and b are constants and $a, b \in R$ and this scheme has randomly generated w_i and is an opportunistic scheme.

2.3.2.3 Adaptive Inertia Weight

$$w_i(t+1) = w(0) + \left[w(n_t) - w(0)\right] \frac{e^{m_i(t)} - 1}{e^{m_i(t)} + 1} \tag{2.8}$$

$$m_i(t) = \frac{g_{\text{best}} - \text{current}}{g_{\text{best}} - \text{current}} \tag{2.9}$$

In an adaptive inertia weight, w_i tends to change depending on the progress of the search and on some factors, which provide information regarding the change in the experience and age of the search agents.

2.3.2.4 Sigmoid Increasing Inertia Weight

$$w_i = \frac{\left(w_{\text{start}} - w_{\text{end}}\right)}{\left(1 + e^{u*(i - n*\text{gen})}\right)} + w_{\text{end}} \tag{2.10}$$

where

$$u = 10^{(\log(\text{gen}) - a)} \tag{2.11}$$

In sigmoid increasing inertia weight, it can be seen that the numerical value inertial weight w_i gradually increases with each iteration, especially with a change in the time factor.

2.3.2.5 Sigmoid Decreasing Inertia Weight

$$w_i = \frac{\left(w_{\text{start}} - w_{\text{end}}\right)}{\left(1 + e^{-u*(i - n*\text{gen})}\right)} + w_{\text{end}} \tag{2.12}$$

where

$$u = 10^{(\log(\text{gen}) - a)}$$

Sigmoid decreasing inertia weight is the counter of the sigmoid increasing inertia weight and here the value of the inertial factor w_i gradually decreases.

2.3.2.6 Linear Decreasing Inertia Weight

$$w_i = w_{\text{max}} - \frac{w_{\text{max}} - w_{\text{min}}}{\text{iter}_{\text{max}}} * i \tag{2.13}$$

In linear decreasing inertia weight w_i also gradually decreases.

2.3.2.7 The Chaotic Inertia Weight

$$w_i = \frac{\left(w_1 - w_2\right) * \text{iter}_{\text{max}} - \text{iter}}{\text{iter}_{\text{max}}} + w * z \tag{2.14}$$

where

$$z = 4 * z * (1 - z) \tag{2.15}$$

In the chaotic inertia weight, the value of the inertia w_i varies nonlinearly and fluctuates without any concrete rule. However, as it follows a definite mathematical expression and some numerical parameter values, hence it is denoted as chaotic and is called as the chaotic inertia weight.

2.3.2.8 Chaotic Random Inertia Weight

$$w_i = 0.5 * \text{rand}() + 0.5 * z \tag{2.16}$$

where

$$z = 4 * z * (1 - z)$$

Unlike the chaotic inertia weight, chaotic random inertia weight is also chaotic and is not random, but the unique feature of chaotic random inertia weight is that it depends on some random numerical value, generated by some random generators. Hence, w_i in chaotic random inertia weight is more fluctuating than the previously discussed chaotic inertia weight.

2.3.2.9 Oscillating Inertia Weight

$$w_i(t) = \frac{w_{min} + w_{max}}{2} + \frac{w_{min} - w_{max}}{2} * \cos\left(\frac{2\pi t}{T}\right) \tag{2.17}$$

where

$$T = \frac{2S_1}{3 + 2i} \tag{2.18}$$

In oscillating inertia weight, the value of inertia w_i is oscillating and depends on some definite fluctuation such as damping or reverse damping or constant oscillating function such as sine or cosine function.

2.3.2.10 Global–Local Best Inertia Weight

$$w_i = \left(1.1 - \frac{g_{best_i}}{(p_{best_i})}\right) \tag{2.19}$$

Global–local best inertia weight has inertial weight w_i based on the value of the fitness function of the individual and the globally best.

2.3.2.11 Simulated Annealing Inertia Weight

$$w_i = w_{min} + (w_{max} - w_{min}) * \lambda^{(k-1)} \tag{2.20}$$

In simulated annealing inertia weight scheme, the inertial weight w_i is generated using the conception and mathematical expression of simulated annealing.

2.3.2.12 Logarithm Decreasing Inertia Weight

$$w_i = w_{min} + (w_{max} - w_{min}) * \log_{10}\left(a + \frac{10t}{T_{max}}\right)$$ (2.21)

In logarithm decreasing where $\lambda = .95$, inertia weight, the inertia w_i depends on a logarithmic value comprising some maximum, minimum inertia value and the information regarding the iteration count.

2.3.2.13 Exponent Decreasing Inertia Weight

$$w_i = (w_{max} - w_{min} - d_1) * \exp\left(\frac{1}{1 + (d_2 t / t_{max})}\right)$$ (2.22)

Here, the inertial weight d_1, d_2 are the control factors to control w between w_{max} and w_{min}. w_i gradually decreases in an exponent decreasing inertia weight scheme.

2.3.3 Fine Grained Inertia Weight Particle Swarm Optimization Algorithm

Fine grained inertia weight particle swarm optimization (FGIWPSO) [6,7] algorithm is yet another variant of the PSO algorithm based on simple nonlinear adaptive strategies, which have a different social influence and guidance scheme for the particles. The novelty of the approach lies in selecting an inertia weight based on a separate mathematical expression. An adaptive inertia weight is involved with each particle of FGIWPSO algorithm, which is tracked by the state space value of the iteration. The selected strategy mainly depends on the numerical value based on the fitness function, which again depends on the spatial value of the position and performance of the particles. Each particle's performance is determined by the absolute difference in the distance of its personal fitness evaluation and the global best evaluation according to the iterations. The inertia weight function gradually fades away exponentially and adaptively with the progress in iteration and is independently applicable for each particle that depends on the iteration or until the particle converges to the global optima.

The basic motive of the PSO algorithm particles is to get influenced by the best positions in the iteration and the global best positions found so far at each iteration, and thus at each moment tries to converge during optimization for the best optima and toward equilibrium state space in case of dynamic environments with the approaching termination criteria of the algorithm.

The role of inertia weight on the contribution of the velocity factor of the previous iteration has an immense contribution and is an important factor in deriving convergence for global optimization. The behavior of PSO algorithm particle's mathematical trajectories clearly indicates that the spatial information of global best particle is the attractor of the whole swarm, and therefore, the movement of each particle is influenced directly by the direction and magnitude of the difference in distance of the global best positional information from the particle's personal best positional information.

In addition, each particle in the PSO algorithm swarm has a different searching capability that is decided by other supporting parameters but is found to be influential and sometimes more decisive than the difference in spatial based influence. Therefore, evaluation of the best inertial value and sometimes a separate and adaptive inertial weight for each particle can converge and better explore the search space. This variation basis can be a good strategy for improving the traditional algorithm. Hence, in FGIWPSO, inertia weight is separately calculated for each of the swarm individuals, and it is based on its performance evaluation at each iteration level.

Mathematically, the variation for the inertial weight w_i for the $(t + 1)$th iteration is visualized by the following relationship:

$$w_i(t+1) = w_i(t) - \left((w_i(t) - 0.4) * e^{-\left(|p_{g\text{best}}(t)| - |p_{i\text{best}}(t)| * \frac{\text{iter}}{\text{iter}_{\max}} \right)} \right) \tag{2.23}$$

The feature of the above-mentioned equation is the involvement of both nonlinear and exponential factors that are dependent on the iteration count and the best spatial information. The participation of the iteration count would naturally ensure that the inertial weight decreases dynamically and adapts with the progress of the particle search.

However, as the search progresses even if the global optimum is approached or not, the inertial weight decreases dynamically to a small value, which facilitates exploitation but hampers exploration resulting in the process of global optimum being stagnated.

2.3.4 Double Exponential Self-Adaptive Inertia Weight Particle Swarm Optimization Algorithm

Depending on the earlier conception and adaptive variation technique another variant of PSO algorithm [7] is being introduced as double exponential self-adaptive inertia weight particle swarm optimization (DESIWPSO) algorithm. It introduces the incorporation of a double exponential *Gompertz function* to select inertia weight for each particle of the PSO algorithm with the progress of iteration. The performance of the particles is framed as a value input to the Gompertz function for determination of inertia weight.

DESIWPSO algorithm is characterized and equipped with the provision to overcome the process of stagnation and the ever decreasing trend of the inertial weight.

The idea in DESIWPSO algorithm is derived from the introduction of the fast-growing nature of double exponential functions over the traditional single exponential, linear, and nonlinear functions. This strategy incorporates a dying double exponential function, that is, the *Gompertz function* for varying the inertia weight of the velocity factor. In general, this function, which is mainly used for modeling of mortality/survival rate of population in a confined space with limited resources, is defined as follows:

$$y = a * \exp\left(b * \exp(x)\right) \tag{2.24}$$

where a is an upper asymptote, and b and x are negative numbers.

Before feeding into the Gompertz function an initial performance index represented by R_i is evaluated for each of the dynamic inertia weight PSO algorithm particles at each iteration based on the particle's personal best position, the global best position of the swarm, and other iteration information belonging to the present iteration value and the maximum possible iteration. The performance index R_i is then fed as an input to the Gompertz function to evaluate the momentum for the weight of the previous velocity value belonging to each particle. The mathematical formulations of R_i and w_i are as follows:

$$R_i(t) = \left| p_{g_{best}}(t) - p_{i_{best}}(t) \right| * \left(\frac{iter_{max} - iter}{iter_{max}} \right)$$ (2.25)

and

$$w_i(t+1) = \exp\left(-\exp\left(-R_i\right)\right)$$ (2.26)

Analysis shows that when the spatial representation of the PSO algorithm particles are far away from the globally best particle in the search space, this DESIWPSO algorithm strategy helps in facilitating better exploration with the help of an adaptive momentum based on the particle's fitness evaluation representing its performance. As particles move toward the optimum or some stable equilibrium on the global best, the inertia is found to facilitate exploitation. Everything depends on the fitness evaluation, but at a later stage the adaptiveness gets stiff, and thus the exploration is minimized, and small steps are considered for movement. This fast-growing nature of the double exponential function increases the convergence speed on iteration but involves some computational processing and time delay due to that computation.

2.3.5 Double Exponential Dynamic Inertia Weight Particle Swarm Optimization Algorithm

Double exponential dynamic inertia weight particle swarm optimization (DEDIWPSO) algorithm is an extension and a second strategy based on the Gompertz function introduced in DESIWPSO algorithm. DEDIWPSO algorithm can be considered as an extended version of DESIWPSO algorithm, and it also incorporates *Gompertz function* to evaluate the inertia weight for each particle depending on the iteration count. Here, the inertia weight is determined for the whole swarm based on a relative increment of iterations for the current iteration. The thought behind such a strategy is to study the behavior of the individuals in the swarm by supplying the same inertia weight to the whole unit irrespective of the count and share of the individual's performance. The resultant behavior of this DEDIWPSO algorithm variant has been better and competitive in performance. This dynamic strategy with the value of inertia weight decreasing iteratively, yet same for all, is an added feature of the DEDIWPSO algorithm as the whole progress of the algorithm approaches the equilibrium state. The governing equations for DEDIWPSO algorithm for selection of inertia weight are as follows:

$$R_i(t) = \left(\frac{iter_{max} - iter}{iter_{max}} \right)$$ (2.27)

$$w_i(t+1) = \exp\left(-\exp\left(-R_i\right)\right)$$ (2.28)

where the parameters and the variables have the same implications as the previous one.

2.3.6 Adaptive Inertia Weight Particle Swarm Optimization Algorithm

Adaptive inertia weight PSO Algorithm is another scheme for varying the steps and thus the search process for the PSO individuals. In this technique, the main governing numerical equation for the inertial weight variation is as follows:

$$w_i = w_{max} - \frac{w_{max} - w_{min}}{iter_{max}} * iter \tag{2.29}$$

where w_{max} and w_{min} represent the range of inertia coefficient, $iter_{max}$ is the maximum number of iterations or generations being allowed or possible, and iter represents the current iteration or generation.

In PSO algorithm, most aspects of the search in the environment are governed by the proper control of global exploration and local exploitation convergence. This is very crucial as far as finding the optimum solution is concerned. In addition, the performance of PSO algorithm individual is highly dependent on the various numerical parameters constituting the mathematical equations.

The first part of the variation velocity equation represents the influence of previous weighted velocity component, which provides the necessary momentum for particles across the search space. The inertia weight denoted by w_i for the ith iteration is the modulus that controls the impact of the previous velocity on the current one. So, the balance between exploration and exploitation in the PSO algorithm is greatly influenced by the value of w_i. Thus, proper control of the inertia weight w_i is found to be very vital for the accurate and efficient determination of the optimum solution. The adaptive inertia weight factor can also be defined as follows:

$$\left[w_i = \begin{cases} w_{min} + \dfrac{w_{max} - w_{max}\left(f - f_{min}\right)}{f_{avg} - f_{min}} & f \le f_{avg} \\ w_{max} & f > f_{avg} \end{cases} \right] \tag{2.30}$$

where we have w_{max} and w_{min} as the defined and estimated but not bounded range for inertial weight w_i and $f()$ is the objective or evaluation function, and correspondingly we have f(avg) as the average and f(min) as the minimum objective values.

It is quite clear that as w_i varies depending on the numerical fitness value of the individual particle, particles with low objective values can be protected, whereas particles with objective values greater than the average will be disrupted. It maintains the legacy that good and better PSO algorithm particles tend to revive and exploit to refine their fitness evaluation by local search, whereas bad particles tend to perform large modifications to explore space with larger steps. Thus, it provides a good way to maintain population diversity and to sustain good convergence capacity.

Another improvement to the PSO algorithm with acceleration coefficients was through a time-varying acceleration coefficient (TVAC) that reduces the cognitive component and increases the social component of acceleration coefficients c_1 and c_2 with time. With a large value of c_1 and a small value of c_2 at the beginning, particles are allowed to move and explore more around the search space, instead of moving toward the present global best particle. A small value of c_1 and a large value of c_2 allows the particles to converge to the global optima during the latter part of the optimization. The TVAC is described by the following mathematical forms:

$$c_1 = \left(c_{1i} - c_{1f}\right)\left(\frac{\text{iter}_{max} - \text{iter}_{min}}{\text{iter}_{max}}\right) + c_{1f} \tag{2.31}$$

$$c_2 = \left(c_{2i} - c_{2f}\right)\left(\frac{\text{iter}_{max} - \text{iter}_{min}}{\text{iter}_{max}}\right) + c_{1f} \tag{2.32}$$

where c_{1i} and c_{2i} are the initial values of the acceleration coefficients c_1 and c_2, and c_{1f} and c_{2f} are the final values of the acceleration coefficient c_1 and c_2, respectively.

2.3.7 Chaotic Inertial Weight Approach in Particle Swarm Optimization Algorithm

Chaotic inertial weight approach (CIWA) in PSO algorithm is yet another version of PSO algorithm having some of the probabilistic features of the Chaotic Sequences in parameter generation in the standard PSO algorithm.

2.3.7.1 Application of Chaotic Sequences in Particle Swarm Optimization Algorithm

First of all, let us see what exactly this chaos theory is. Chaos is the apparent disordered unique behavior that is deterministic but is a universal phenomenon found to occur in many scientific experiments. The applications of the chaotic sequences have been proven to be very efficient and have produced promising results in many engineering applications. The basic equation of the chaotic variation is given by the following equation:

$$\gamma_k = \mu * \gamma_{k-1} * \left(1 - \gamma_{k-1}\right) \tag{2.33}$$

where μ is a control parameter and $\mu \in R$, and γ_k is the chaotic parameter at the corresponding iteration k.

The behavior of the swarm is greatly influenced by the variation of μ, and its value determines whether γ is stabilized at a constant size, oscillates between limited sequences of sizes, or behaves chaotically in an unpredictable pattern. The performance of a PSO can depend on its parameters, that is, the inertia weight factors and two acceleration coefficients. The first term in the velocity update represents the influence of the previous velocity multiplied by the inertia, which provides the necessary momentum for particles to fly around in a search space. The balance between exploration and exploitation can be treated by the value of inertia weight.

To improve the global optimization search capability and to increase the probability of escaping from a local minimum, a new weight-changing approach, CIWA, is suggested as follows:

$$w_{ck} = w_k * \gamma_k \tag{2.34}$$

where, w_{ck} is a chaotic weight at iteration k, whereas w_k is the weight factor from the inertial weight approach, and γ_k is the chaotic parameter. The weight in the conventional inertial weight approach decreases monotonously and varies between $w_{\max}$ and $w_{\min}$, whereas the chaotic weight decreases and oscillates simultaneously [8]. The CIWA approach can encompass the whole weight domain under the decreasing line in a chaotic manner; the searching capability of the proposed algorithm can be increased.

2.3.7.2 Crossover Operation

Crossover operator in the PSO algorithm introduces more variation for the particles, and thus provides better search for work space. This kind of diversity not only helps in an effective exploration but also helps in exploitation, and the mathematical representation for crossover operation is provided as follows:

$$x_{ij}^{(k+1)} = \begin{cases} x_{ij}^{(k+1)} & \text{if } r_{ij} \le \text{CR} \\ x_{ij}^{(P,k)} & \text{otherwise} \end{cases} \tag{2.35}$$

where $x_{ij}^{(k+1)} \in X_{ij}^{(k+1)} = (x_{i1}^{(k+1)}, x_{i2}^{(k+1)}, \ldots, x^{(k+1)})$ is formed from $X_i = (x_{i1}, x_{i2}, \ldots, x)$ and the global best particle $X_{(P,i)} = (x_{i1}^{(P)}, x_{i2}^{(P)}, \ldots, x^{(P)}).j \in (1, 2, \ldots, n)$ where r_{ij} is a uniformly distributed random number, and CR is the crossover rate. When the value of CR becomes one, there is no crossover similar to in the conventional PSO algorithm. If the value of CR is zero, the position will always have the crossover operation similar to the genetic algorithm mechanism. A proper crossover rate CR can be determined by empirical studies in order to improve the diversity of a population.

2.3.8 Distance-Based Locally Informed Particle Swarm Optimization Algorithm

These groups of PSO particle mainly gets influenced by the positional information of the other particles and have come out the traditional thought process that only global best can influence better search and convergence.

2.3.8.1 Fitness Euclidean-Distance Ratio Particle Swarm Optimization

Fitness Euclidean-distance ratio PSO (FERPSO) [9] is a variant of the PSO in which the traditional criteria of getting influenced by the global best position are being discarded, and instead the best nearest or a neighboring particle is being selected for better exploration. The criteria will, no doubt, enhance the search but at the cost of increasing work for the tracking system, which will constantly monitor the nearest neighbor for each particle. However, for the sake of performance, mission-critical applications may choose it to be traded off. So, the algorithm boils down to the social influence of the personal best on the fittest and closest neighbors (n_{best}), which are identified by the fitness—Euclidean distance ratio (FER) values. The n_{best} for ith particle is selected as the neighborhood personal best with the largest FER as follows:

$$\text{FER}_{(i,j)} = \alpha . \frac{f_{(p_j)} - f_{(p_i)}}{|p_j - p_i|} \tag{2.36}$$

where $\alpha = \|s\| / f(p_{\text{gas}}) - f(p_w)$ is a scaling factor. It can be seen that the value of α can scale the result to some particle other than the nearest best particle. In addition, we have p_w

and p_{gas} as the worst and global best-fitted PSO particle, whereas p_j and p_i are the personal best for the jth and ith particle. $\|s\|$ is the size of the search space estimated by the diagonal distance $\sqrt{\sum_{k=1}^{d}(x_k^u - x_k^l)^2}$ where the x_k^u and x_k^l are the upper and lower bounds of the kth dimension search space. The introduction of the search space capacity will decrease the number of out of range solutions and will try to stick to the bounds. This will help in better exploration and will result in a decrease in the number of discarded solutions. The final influencing factor or velocity for the particles can be represented by the following equation:

$$V_i^d = \omega * V_i^d + c_1 * \text{rand}1_i^d * \left(p_{best i}{}^d - X_i^d\right) + c_2 * \text{rand}2_i^d * \left(n_{best}{}^d - x_i^d\right) \qquad (2.37)$$

where $\text{rand}1_i^d$ and $\text{rand}2_i^d$ are two random numbers within the range of [0,1]. We have ω as the inertia factor, p_{best} is the personal best, and n_{best} is the next neighbor, which is having a considerable fitness evaluation. The rest of the notations have their usual meanings. However, it can be seen that this next door neighbor concept will not only help in a better local search and exploration but also prevent particles from getting accumulated at a certain place and prevent the unimodal behavior. Thus, it can help in multimodal functions and mainly for dynamic systems in which the optimum moves constantly, and there is a requirement for better exploration.

2.3.9 Speciation-Based Particle Swarm Optimization

Speciation-Based PSO (SPSO) was framed in 2004 [10]. The main specialty of this kind of the PSO is the subgrouping of the particles in the form of species and thus distributing the social influence for each group in the form of different global best for each group. For each species, there are definite influencers called species seeds. However to determine the species seeds and the size of each species, some regional information must be provided by the search space in the form of coverage area or radius. The species seed and its group work cooperatively but independent of the other group, and the whole system consisting of several species constitutes the overall PSO system and thus produces the required optimization. SPSO, unlike FERPSO, replaces the global best by species best or species seed and all the particles of the same species share the same neighborhood best.

2.3.10 Ring Topology Particle Swarm Optimization

Ring topology PSO is another variant of PSO in which the particle interacts with its neighbors as its social mates and gets influenced. Thus, there forms a chain of interacting particles and this kind of multi-niche influence has been realized to be efficient in multimodal optimization. PSO algorithms using the ring topology are successful in forming stable niches across different local neighborhoods, eventually locating global or multiple local optima.

2.3.11 Distance-Based Locally Informed Particle Swarm

Distance-based locally informed particle swarm [11] was introduced to overcome the disadvantages of the traditional PSO for polarization of the particles for single optima. The less credibility of traditional PSO for multimodal optimization was because all the particles tend to get influenced by a single best particle position and this is the reason that exploration is not complete. Multimodal optimization problems are characterized by topology-based separated and distributed optima and they can be very far from one another mainly when the search space is very large, and the step size in comparison is very low. In addition, for the neighbor-based influenced PSO the particle had the same problem of getting converged to one point, which is, perhaps, not desired. For instance, in ring topology, the particles are allowed to get influenced by their neighbors and get polarized to them, instead of trying to reach the next unsearched area of interest in which better solutions might exist. This kind of local struck is prevalent as these particles behave on the influence of the coparticles.

The success of PSO for this kind of problem is highly dependent on the initialization, which is also a stochastic process and depends on the randomness of the generator. If the neighbors are properly placed, then the algorithms will find it very difficult to converge for multiple solutions and move toward other located peaks. Distance-based locally informed particle swarm uses the following velocity update equation for positional equation X so that it can incorporate several influences of the best particles:

$$V_i^d = \omega \times \left(V_i^d + \varphi \left(p_i^d - X_i^d \right) \right) \tag{2.38}$$

where

$$p_i = \frac{\sum_{j=1}^{\text{nsize}} \left(\varphi_j . n_{\text{best}_j} \right) / \text{nsize}}{\varphi} \tag{2.39}$$

Here, we have φ_j as a uniformly distributed random number in the range of [0, 4.1/nsize] and φ is equal to the summation of φ_j, that is, $\varphi = \sum \varphi_j . n_{\text{best}_j}$ is the jth nearest neighborhood to the ith particle's p_{best}, nsize is the neighborhood size.

Locally informed particle swarm has the advantage of full utilization of the velocity factor for high exploration of the search space through the usage of the neighboring information and also Euclidean distance-based neighborhood selection will ensure the algorithm's ability for local search and fine-tuning through the selection of the influencing factor from the next door neighbor. The neighborhood selection ensures that there is a clear demarcation of different optima regions in the search space and thus helps in a better multimodal optimization search.

There are several advantages of this kind of Euclidean distance-based neighborhood selection over the traditional best topological neighborhood influence. This is because in the case of the multimodal optimization the requirement is more toward what the best neighbor has and the search is directive instead of biased. In addition, the difference in the direction and difference in neighbor will create a difference in what we called exploration and it will not make the particles converge toward some end. We can easily see that the lesser difference in the distance between the two particles will gradually destroy the urge for movement for the particles and the overall exploration will not be there. Another criterion that must be there is that the particles must scatter to every nook of the search space, and this will satisfy the search criteria of implementing so many particles. However, unlike natural PSO algorithm in which there is an influencing factor in each case, irrespective of the search that has been taken place and the success it has received, there may or may not be topological neighbors from different niches and this case cannot be avoided at any cost for the particles. The resultant effect will be several oscillations between two peaks, and that can waste a number of searchless oscillations. The main reason for searchless oscillations is the position of the two particles at the two niches, and this condition is detrimental for the mixture of an appropriate exploration with exploitation. The particles are more concentrated to locate additional niches for a multimodal situation during the initial exploration and then try to avoid the oscillation between the two niches in which it has got stuck. Basically, locally informed particle swarm will be able to perform a search within each Euclidean distance-based neighborhood without any interference from another niche present in other parts of the search sector. Locally informed particle swarm is highly effective for fine search as it is successful in locating the desired peaks with a high accuracy. That is why locally informed particle swarm will perform better than the traditional PSO algorithm.

2.3.12 Inertia-Adaptive Particle Swarm Optimization Algorithm with Particle Mobility Factor

Inertia-adaptive PSO algorithm is a premiere version of the PSO algorithm in which the inertia ω of the velocity factor is made a variable, which gradually changes with iterations and creates a new stir in the motion of the particles, and thus helps in better exploration. The main reason for introduction of this kind of adaptive inertia is the premature convergence, which can occur when the intradistance of the particles and the global best is very less, and there is very less change in the topological information of the best position considered so far. This can not only cause less exploration and decrease in reachability of the particles to every corner of the search space but also can reach the near optimal solution. Thus, the real optima will not be reached and the tendency of the influence will lead the particles astray. Thus, the following equation represents the velocity equation:

$$v_i(t) = \omega v_i(t-1) + r_1 c_1 \left(x_{ib} - x_i(t-1) \right) + r_2 c_2 \left(x_{gb} - x_i(t-1) \right) \tag{2.40}$$

We have value of ω for each particle given by

$$\omega = \omega_0 \left(1 - \frac{\text{dist}_i}{\text{max}_{\text{dist}}} \right) \tag{2.41}$$

The first of these modifications involves modulation of the inertia factor ω according to a distance of the particles of a particular generation from the global best, where $\omega_0 = \text{rand}(0.5,1)$, dist_i is the current Euclidean distance of ith particle from the global best, that is,

$$\text{dist}_i = \left(\sum_{d=1}^{D} \left(g_{\text{best}_d} - x_{i,d} \right)^2 \right)^{1/2} \tag{2.42}$$

and max_dist is the maximum distance of a particle from the global best in that generation, that is,

$$\text{max\_dist} = \text{argmax}_i(\text{dist}_i)$$

The main reason for introducing this variable factor is that if the absolute value of the supporting terms such as $\left| p_{\text{best}i,d} - x_{i,d} \right|$, $\left| g_{\text{best}_d} - x_{i,d} \right|$ are very small and $v_{i,d}$ for the previous iteration is very less then it may happen that the all other supporting factors such as ω, *r*, and *c* can create very less influence for the displacement and will constraint the phenomenon of exploration. For this reason, what we need is high gravity inertia for the motion of the particles, and this can be achieved only through adaptive inertia, which senses the situation or even can change itself randomly for experimentation of the situation for the particles. Thus, even we compromise in the initial phase of the exploration with less inertia, with time we need to switch to some better ones in the later phases, and this must be adaptive and automatic for the situation, condition, and bound of the search space of the problem. This loss of the diversity and the ability to create a spreading effect among the particles is a serious drawback of the PSO algorithm.

The main reason for the modulation index in the inertia term is to help the PSO particles in exploration by making it to move away from the global best. This is ensured by the factor that the attraction toward the global best will predominate, and premature convergence can be avoided. This is done through the release from local optima toward a better position, which is yet to appear. This will be beneficial for the multimodal landscapes, which represent several real-life problems from now onward, we shall refer to this new algorithm as inertia-adaptive PSO (IAPSO).

2.3.13 Discrete Particle Swarm Optimization Algorithm

Discrete particle swarm optimization (DPSO) algorithm is another intelligent behavior of the algorithm for some special discrete event-based applications or rather the applications whose solution can be chosen from fixed sets of zeros and ones.

The equations governing the DPSO algorithm are as follows (which mainly includes one extra equation and a bit modified implications):

$$x_{i(t)} = x_{i(t-1)} + v_{i(t)} \tag{2.43}$$

$$v_i(t) = w v_i(t-1) + r_1 c_1 \left(x_{ib} - x_i(t-1) \right) + r_2 c_2 \left(x_{gb} - x_i(t-1) \right) \tag{2.44}$$

Considering a threshold Th, we consider another set of equations as follows to convert the continuous values into the discrete form. Here, we consider the binary values that are either 1s or 0s and in the case of the traveling salesman problem (TSP), the binary sequence is converted into decimal for generating the sequence of the nodes that need to be covered for optimal or less cost operation.

The binary sequence is denoted as x_i^b as $\{x_1^b, x_2^b, \ldots, x_n^b\}$ as transformed from $\{x_1, x_2, \ldots, x_n\}$

$$x_{ij}^b = \begin{cases} 1 & \text{if } V_{ij} > \text{Threshold} \\ 0 & \text{otherwise} \end{cases} \tag{2.45}$$

This can also be depicted as

$$x_{ij}^b = \begin{cases} 1 & \text{if } V(0,1) < \text{sigmoid}(V_{ij}^k) \\ 0 & \text{otherwise} \end{cases} \tag{2.46}$$

where $V(0,1)$ function is $V(0,1) \approx [0,1]$ and sigmoid function is the mapping from the real value to the [0,1] range.

DPSO algorithm has been utilized in the solution of TSP and QAP in which each of the agents carries a discrete value following the traditional equations of the DPSO algorithm.

Steps for the Discrete Particle Swarm Optimization Algorithm

Procedure (Discrete PSO){graph, PSO-agents, globaloutput}
Initialize N Agents with D dimension represented as $\{x_1, x_2, ..., x_D\}$
Conversion: Problem Model as a Graph Problem $G(V,E)$
Initialization Graph Matrix, PSO-Agents, DPSO-Agent,
Initialization Agent Memory, and Initialize to random value
For {PSO-Agents = 1 to N}
For {PSO-AgentsSubset{i} with i = 1 to M}
  If PSO-AgentsSubset{i,j} > Threshold
   DPSO-AgentSubset{i,j} = 1
  Else
   DPSO-AgentSubset{i,j} = 0
EndIf
EndFor
EndFor
Calculate Fitness of DPSO-Agents
if Best then (N Agents) is.BetterGlobal.Best
  Update Global.Best
end if
for Iteration i = 1 to Iteration$_{max}$ do
for PSO-Agents j = 1 to N do
for PSO-Agents Dimension k = 1 to D do
Calculate r_1, c_1, r_2, c_2
Calculate $v_{jk}(i) = w * v_{jk}(i-1) + r_1 c_1 (x_{gb}(i-1) - x_{jk}(i-1)) + r_2 c_2 (x_{ib}(i-1) - x_{jk}(i-1))$
Calculate $x_{jk}(i) = x_{jk}(i-1) + v_{jk}(i)$
end for
end for
For {PSO-Agents = 1 to N}
For {PSO-AgentsSubset{i} with i = 1 to M}
  If PSO-AgentsSubset{i,j} > Threshold
  DPSO-AgentSubset{i,j} = 1
  Else
  DPSO-AgentSubset{i,j} = 0
EndIf
EndFor
EndFor
Calculate Fitness of DPSO-Agents
If Best then (N Agents) is.BetterGlobal.Best
  Update Global.Best
end if
end for
end procedure

2.3.14 Particle Swarm Optimization Algorithm for Continuous Applications

PSO algorithm has been used in various kinds of applications for numerical optimization. One such example is being discussed here. Equation of circle for optimized global minima at (0,0). Search area is defined as $-15 \leq x, y \leq 15$. Let the four particles are initialized as follows:

We consider $c_1 = c_2 = 2$ and $r_1, r_2 = [0,1]$, $w = 1.2$
Particle 1: $x = (2.8, -3.1)$ fitness $= 4.18$, $v = (0.9, 0.2)$
Particle 2: $x = (-1.9, -2.1)$ fitness $= 2.83$, $v = (-0.72, 0.43)$
Particle 3: $x = (-0.8, 2.3)$ fitness $= 2.44$, $v = (0.9, -1.8)$
Particle 4: $x = (4.2, 1.74)$ fitness $= 4.55$, $v = (-0.67, 0.28)$

Initially at 0th iteration Global Best = Iteration Best = $(-0.8, 2.3)$
But we assume and consider that Global Best = $(0.3, -0.2)$ and Iteration Best = $(-0.8, 2.3)$
For the first iteration, we calculate the value of v and x for each particle as

Particle 1: $v = ((0.9 * 1.2 + 2 * .5 * (2.8 - (-0.8)) + 2 * .3 * (2.8 - 0.3)),$
$(0.2 * 1.2 + 2 * .4 * (-3.1 - 2.3) + 2 * .2 * (-3.1 - (-0.2)))) = (6.18, -5.24)$
$x = ((2.8 + 6.18), (-3.1 + (-5.24))) = (8.98, -8.34)$
Fitness $= 12.26$

Particle 2: $v = (-0.72 * 1.2 + 2 * .35 * (-1.9 - (-0.8)) + 2 * .75 * (-1.9 - 0.3)),$
$(0.43 * 1.2 + 2 * .15 * (-2.1 - 2.3) + 2 * .53 * (-2.1 - (-0.2)))) = (-4.93, -2.82)$
$x = ((-1.9 + (-4.93)), (-2.1 + (-2.82))) = (6.83, 4.92)$
Fitness $= 8.41$

Particle 3: $v = ((0.9 * 1.2 + 2 * .71 * (-0.8 - (-0.8)) + 2 * .25 * (-0.8 - 0.3)),$
$(-1.8 * 1.2 + 2 * .19 * (2.3 - 2.3) + 2 * .5 * (2.3 - (-0.2)))) = (0.53, 0.34)$
$x = ((-0.8 + 0.53), (2.3 + 0.34)) = (-0.27, 2.64)$
Fitness $= 2.65$

Particle 4: $v = ((-0.67 * 1.2 + 2 * .41 * (4.2 - (-0.8)) + 2 * .57 * (4.2 - 0.3)),$
$(0.28 * 1.2 + 2 * .26 * (1.74 - 2.3) + 2 * .55 * (1.74 - (-0.2)))) = (7.74, 2.18)$
$x = ((4.2 + 7.74), (1.74 + 2.18)) = (11.94, 3.92)$
Fitness $= 46.80$

Global Best = $(0.3, -0.2)$

Iteration Best = $(-0.27, 2.64)$ as the last iteration has this as the smallest.

The whole process continues until the stopping criteria are met. There is a high probability that the solution may get better or worse and depending on the search space and the position of the fellow particles. The combinatorial influence may lead the particle to spaces in which the solution is not favorable, but that forms the part of the search analysis. A brief discussion on convergence is given in Section 2.4.

2.4 Convergence Analysis of Particle Swarm Optimization Algorithm

PSO algorithm has high convergence rate so far as swarm intelligence is concerned. The main convergence is because of the social influence from the best particle in the lot. The social influence can induce movement in both (toward and away from the global best) the directions. However, the dynamics of the PSO algorithm changes due to the constant change in the global optima and exploration. PSO algorithm has another two parameters, r and c, which can also help in an uneven movement in the search space and the overall vector summation can help in exploration. However, in this regard, it can be said that, with the progress of the iteration there is a chance of convergence of the particles. Convergence of the particles means that the particles tend to accumulate at a certain position mainly near the global best. In that situation, the difference between the particles tends to fade away and they create negligible influence for the fellow ones. The exploration gradually gets hampered, and the unvisited portions remain neglected. This is not desirable as optima can remain distributed at any position of the search space. The exploration can be regained through reinitialization. The variants of the PSO algorithm have overcome some of these problems and come with an enhanced architecture for exploration and exploitation.

2.5 Search Capability of Particle Swarm Optimization Algorithm

The main reason for the world-wide acceptance of the PSO algorithm is its searching capability. If we look at the dynamics of the particle and its movement, we can see that the particle is getting influenced by two factors, one is the global best and other is the iteration best or personal best. Overall, we can say that the influence is the vector sum of the two. The way of movement of the particle toward the global best is the algebraic sum of the positional influences and the movement can be either toward or away from the global best. The directional value of their algebraic sum plays the key role in such a situation. But what else are these two equations producing? The PSO algorithm has a constant search movement and in that case unless they come across the real global optima their global best is changing constantly, and so is the iteration best and this is the reason why there is so much scattering of the particles in the form of exploration.

2.6 Summary

In this chapter, we have discussed some important variants of the PSO algorithm along with the traditional one and have clearly revealed how the PSO algorithm gradually evolved from its basic form to the complex mathematical model demanding high computation but providing better search and exploration in the workspace and a better convergence rate. High rate of exploration will provide better dynamics for the particles, will not restrict them to the local optima, and will break out of the global best through the diversified variation and decision-making. PSO algorithm is one of the most successful and basic kind of optimization technique and is easy to understand the conception of particles. The introduction of the variations has helped and taught us how playing with the mathematical complexities can derive a better result. Nevertheless, some of things such as parameter initialization for the search space and so on, are beyond discussion and require some monitoring and experimentation and vary highly from case-to-case depending on the situation and data bound of the considered problem. So, we have covered the following characteristics and description of the PSO algorithm in this chapter:

- The traditional PSO algorithm
- The variants of the PSO algorithm that have better performance and enhanced features for exploration
- The variants of the PSO algorithm such as sequential PSO algorithm, inertia weight strategies in PSO algorithm, FGIWPSO algorithm, DESIWPSO algorithm, DEDIWPSO algorithm, adaptive inertia weight particle swarm optimization algorithm, and CIWA in a PSO algorithm
- The discrete version of the PSO algorithm that can be utilized for certain special graph-based applications
- The discussion related to the convergence and dynamics of the PSO algorithm and how the different variants are proven to be better
- A few examples to show how the PSO algorithm is working both with respect to continuous and discrete problems

Solved Examples

1. What is the significance of Gait in the identification and recognition of Parkinson's disease? How PSO can be used for the recognition of Parkinson's disease using Gait?

Solution 1

Parkinson's disease is a neurological disorder that progresses with time. People suffering from Parkinson's disease lack dopamine, which is a chemical present in nerve cells of the brain. Without this chemical, the movement becomes slower in people suffering from this disease. The main symptoms of Parkinson's are tremor, rigidity, and slowness of movement. The disease does not directly cause people to die, but symptoms get worse as the time progresses.

Therefore, the early detection and severity staging of Parkinson's disease have become a very important issue. This assessment of Parkinson's disease can be done with the identification of specific gait patterns.

Gait is an emerging biometric technology field to recognize a person purely by analyzing the way they walk. Gait has some unique characteristics. The most attractive feature of gait as a biometric trait is unobtrusiveness. In other words, unlike other biometrics, gait can be captured at a distance and without requiring prior consent of the observed subject. One advantage with gait is that it can be detected and measured at low resolution. It also has the advantage of being difficult to hide, steal, or fake.

Gait recognition method can be classified into two main classes:

1. *Model-based method*

    Model-based methods fit a model to the image data and then analyze the variation of its parameters. The majority of the current model-based approaches are based on 2D methods. It always assumes an explicit model of the body structure or the motion and then performs model matching frame by frame. The effectiveness of these techniques is limited, for example, recovering parameters or modeling human body from walking sequence has been a challenging task.

2. *Model free method*

    Model-free method extracts statistical features from a subject's silhouette to distinguish walkers. The statistical method does not need to construct any prior model initially. Rather, it extracts some statistical parameters directly from an image sequence. The sequence may include a set of walking poses captures from some object.

    The methods employed for the detection of disease extract too many features from the subject. So, PSO algorithm can be used to reduce the feature set and employ proper soft computing techniques to train the dataset, which can help in the identification of affected patients. The reduced feature set will also decrease the lab cost for the diagnosis of Parkinson's disease (Detailed application in given in solved example 2).

2. How can we optimize the features in *gait* database using the PSO algorithm? Illustrate considering the benchmark datasets.

Solution 2

Illustration of Particle Swarm Optimization Application

PSO optimization technique is applied over the two gait datasets of Parkinson patients. The purpose of PSO algorithm is to optimize or reduce the required features for the recognition of Parkinson disease. As the class to which dataset belongs is known beforehand, we appended specific bits to indicate the category of the instance. To each instance that belongs to the patient affected with Parkinson disease, 1 is appended. The other instances are appended with 0 marking the unaffected patients.

The entire experiment of feature selection was carried out in the Eclipse integrated development environment (IDE). The dataset is of physionet gait dataset. The PSO algorithm is then applied over the dataset and the relevant features are extracted. Nineteen-feature dataset is reduced to five-feature dataset using the PSO algorithm (Figures 2.1 through 2.3). The selected feature subset obtained from the PSO algorithm is applied on the database and the accuracy (with original dataset) is summarized in Table 2.1.

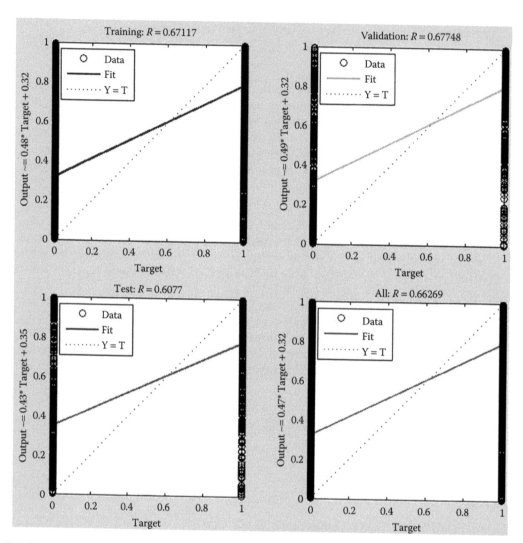

FIGURE 2.1
Regression plot of Ga dataset (PSO Algorithm).

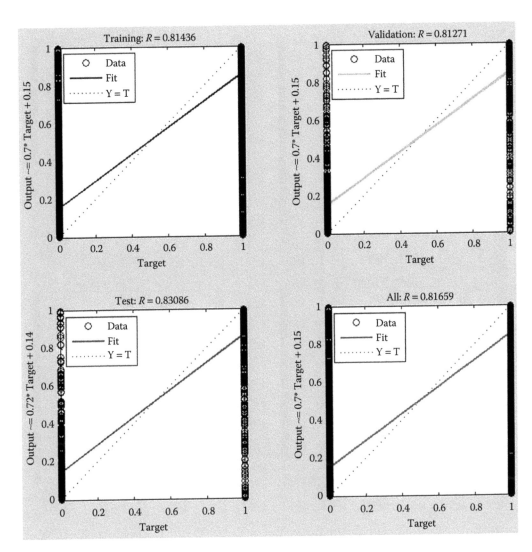

FIGURE 2.2
Regression plot of Ju dataset (PSO Algorithm).

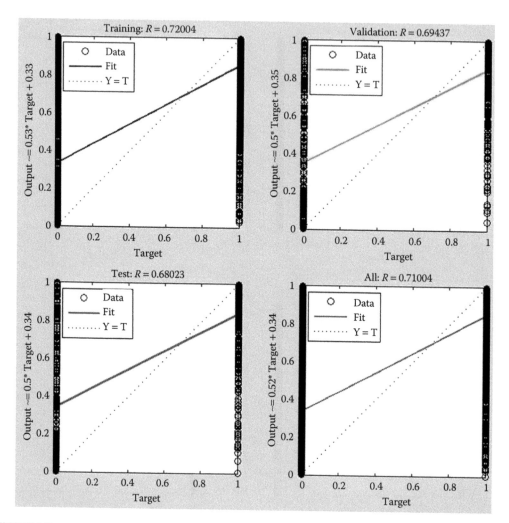

FIGURE 2.3
Regression plot of Si dataset (PSO Algorithm).

TABLE 2.1

Classification Accuracy of PSO Algorithm over an Optimized
Physionet Dataset

Algorithm	Dataset Name	Accuracy (%)
PSO	Ga	67.74
	Ju	81.27
	Si	69.43

Exercises

1. Implement the basic PSO algorithm in the programming language of your choice.
2. Discuss how certain variants of PSO prove to be advantageous over the primary one in certain applications.
3. What is the role of *best position* in the PSO algorithm?
4. Why is acceleration coefficient required in PSO algorithm? Can it be compensated with the velocity component? Justify your answer.
5. Discuss the role of benchmark functions in any artificial intelligence algorithm.
6. Under what conditions, can the original PSO algorithm fail to converge?
7. Discuss the phenomenon of *premature convergence*. How can it be avoided?
8. What is the physical significance of *inertia weight*?
9. Why was there a need to introduce *constriction coefficient* in the velocity update rule?
10. What is meant by *constraint optimization*? How is it different from *optimization* implemented in the PSO algorithm?
11. How should an original PSO tune itself to function in a dynamic environment?
12. TSP: 10 cities, namely, A–J alphabetically, are located at the Cartesian locations (10,0), (20,5), (35,10), (40,15), (25,20), (25,30), (20,30), (10,20), (5,5), and (25,10), respectively.
    a. Solve the above-mentioned TSP for particle count = 8 if maximum velocity changes are bound to 5. You may choose a convenient upper bound for the maximum number of iterations for the algorithm.
    b. Solve the above-mentioned TSP for particle count = 4 if maximum velocity changes are bound to 10. You may choose a convenient upper bound for the maximum number of iterations for the algorithm.
    c. Comment on the performance of the algorithm in the above-mentioned two cases, that is, 12a and 12b. Is there an optimum value of particle count and maximum velocity changes allowed, which can result in an optimum performance in the given case?
13. *Pattern searching problem*: Implement PSO to search for a target word = SWARM. Let the range of characters to be searched be letters from *a* to *z*. Let MAX = maximum number of letters that can be put together, MIN = minimum number of letters that can be put together, COUNT = number of particles to use, V_MAX = maximum velocity allowed, and V_MIN = minimum velocity allowed.
    a. Solve the above-mentioned problem for MAX = 7, MIN = 3, COUNT = 8, V_MAX = 5, and V_MIN = 2. Assume a convenient number of iterations.
    b. Solve the above-mentioned problem for MAX = 10, MIN = 1, COUNT = 25, V_MAX = 4, and V_MIN = 1. Assume a convenient number of iterations.
    c. Comment on the performance of the algorithm in the above-mentioned two cases, that is, 13a and 13b. Does increasing the particle count aid the algorithm? What should be the values of MAX, MIN, and COUNT for optimum performance? Is it intuitive to provide such values in a problem in which the target is less obvious?

References

1. J. J. Liang, A. K. Qin, P. N. Suganthan and S. Baskar, Comprehensive learning particle swarm optimizer for global optimization of multimodal functions, *IEEE Transactions on Evolutionary Computation*, 2006, 10(3), 281–295.

2. J. Kennedy and R. Eberhart, Particle swarm optimization, *Proceedings of ICNN'95—International Conference on Neural Networks*, 1995, 4, 1942–1948.

3. A. Sallama, M. Abbod and S. M. Khan, Applying sequential particle swarm optimization algorithm to improve power generation quality, *International Journal of Engineering and Technology Innovation*, 2014, 4, 223–233.

4. X. Zhang, W. Hu, S. Maybank, X. Li and M. Zhu, Sequential particle swarm optimization for visual tracking, *Proceedings of the 26th IEEE Conference on Computer Vision and Pattern Recognition (CVPR '08)*, 2008, pp. 1–27.

5. J. C. Bansal, P. K. Singh, M. Saraswat, A. Verma, S. S. Jadon and A. Abraham, Inertia weight strategies in particle swarm, *Proceedings of Third World Congress on Nature and Biologically Inspired Computing*, 2011, pp. 640–647.

6. K. Deep and P. Chauhan, A new fine grained inertia weight particle swarm optimization, *2011 World Congress on Information and Communication Technologies (WICT)*, 2011, pp. 424–429.

7. P. Chauhan, K. Deep and M. Pant, Novel inertia weight strategies for particle swarm optimization, *Memetic Computing*, 2013, 5, 229–251.

8. J. Park, Y. Jeong, J. Shin, S. Member and K. Y. Lee, An improved particle swarm optimization for nonconvex economic dispatch problems, *IEEE Transactions on Power System*, 2010, 25(1), 156–166.

9. J. J. Liang, B. Y. Qu, S. T. Ma and P. N. Suganthan, Memetic fitness euclidean-distance particle swarm optimization for multi-modal optimization, *Bio-Inspired Computing and Applications*, 2012, pp. 378–385.

10. S. Bird and X. Li, Enhancing the robustness of a speciation-based PSO, *2006 IEEE Congress on Evolutionary Computation*, Vancouver, Canada, July 16–21, 2006, Vol. 1, no. 1, pp. 843–850.

11. B. Y. Qu, P. N. Suganthan, S. Member and S. Das, A distance-based locally informed particle swarm model for multimodal optimization, *IEEE Transactions on Evolutionary Computation*, 2013, 17(3), 387–402.

3

Genetic Algorithms

3.1 Introduction

This chapter will deal with some of the most widely used and successful evolutionary algorithms, which are inspired from the evolution of natural evolution and organisms. They are mostly used in discrete problem optimization but can also be used in multi-objective continuous problems as well. The main advantages of using these algorithms are its combination generation, learning capability, and using the learnt experience for a generation of better solutions. There are various evolutionary algorithm like genetic algorithms (GA), genetic programming approaches, and evolutionary strategies [1–4]. There are various modified versions of these algorithms, which have been introduced when certain deficiencies were found in the original ones. In addition, there were times when it was found that the modified versions were far better in performing than the traditional algorithm itself. We will gradually get into the details of the algorithm and will illustrate with examples so that the readers can visualize the operations and can use them in their problems for optimization and solution generation. However, the explanations provided are just one-sided visualizations, and there can be several other observations and techniques of using the algorithm for the various problems and applications [5,6].

GA is regarded as one of the pioneers of soft computing techniques and the oldest probabilistic meta-heuristic [7,8]. Apart from the neural network, which was a learning-capable procedure and a fuzzy logic, which was for regression and combined decision-making, GA emerged as the only meta-heuristics [9], which was able to derive solutions through combination and developed them with iteration, and thus provided optimization depending on its previously learnt experience. If the details of the GA are studied and analyzed on how it has been able to set up the platform for solutions, it can be seen that the learning capability of the GA is normally passive and not active as similar in a neural network. This passive learning can be achieved through evaluation of the temporary solutions and their subset. Later, the same learnt experience is used for the proper development and enhancement of the solutions. This phenomenon is very much similar to the development of species through the combination of chromosomes. However, it has been seen that the species with suitable chromosomes only survived and also the combination does not follow any criteria [10–12]. Chromosomes in real life are combined with opportunity and randomly without preestimation. The GA thus formed follows some procedure, and it has been seen that each one of them follows some procedure and has some advantages and disadvantages. Sometimes the best possible outcome is achieved through the involvement of all of them but tends to consume some more processing power, which can be spent for better convergence. The details of the arguments are discussed later in the chapter when we take up the algorithm and its various operations and subcomponents. This marked the end of the era of traditional thinking that algorithms are deterministic procedures and must produce the same result(s) in a limited number of steps or iterations no matter how many times it has been run. Deterministic algorithms have the traditional thinking for procedures. The operations tend to get biased on the local minima such as taking the least weighted path and so on. However, such greediness seldom has achieved optimization or rather the least weighted route to the destination. GA was first introduced by Holland in the year 1975 based on the theories of natural phenomenon given by Darwin and later nurtured by Mendel. GA has broken the shackle for traditional thinking that algorithm will always produce the same result no matter how many times it has been executed. Dynamics, convergence, and stability, have been the new criteria for assessment of the algorithms instead of complexity and other consumption stuff. This is because people are trying to solve a bigger problem that involves many parameters and each of them behaves independently. Hence, it is not worthy to follow-up each of them individually but as a group, and the group behavior will throw light on the individual collectively. After some initial years before it was assimilated by the researchers as an accepted procedure for solution generation, it went past many established algorithms not only through its solution generation capability, but also because of its convergence and optimization capability [13–15]. It had been most successful in those kinds of problems in which there exist several valid solutions but only limited optimized and acceptable solutions. Combinatorial optimization problems are one kind of such problems. The challenge in such problems is that the algorithm cannot determine its stopping criteria and movement, and hence stochastic exploration is the only way out for solution derivation. Before getting into the procedures for several applications of GA let us investigate the reality and inner capability of the individual procedures of the GA.

GA has some critical criteria for its operation and sometimes demands situations for reliable optimization. One such condition is a Non-dynamic objective function. Dynamical problems can be traced and roughly predicted but cannot be optimized because there is no stability of the solution. Stability of the solutions can only be achieved when the system is frozen in time. However for dynamic systems, the solution generation time must be less than the change in dynamics of the system and such assurance is hard to achieve. Traditional algorithms have been evaluated with the complexity of time and space. However, GA has always been found to be memory hungry. This is because of the generation of the chromosomes in a bucket known as the gene pool. The quality of the gene pool had been very vital in many cases. More important is the combination of the chromosomes in the gene pool. Individual chromosomes, if considered, and its fitness are essential when solution quality or optimization is concerned, but the motive of GA is a combination generation in new possible ways, and these indulge the gene pool to be multicombined that consists of different individuals with respect to the fitness. The success of the GA will never rely on the operations but on the individual chromosomes and sometimes on the opportunity created. GA has been regarded as an algorithm, which operates on nonlocal search because of the frequent crossover operation [16,17] and mutation. Hence, convergence is subjected to the generated solutions based on the operations and is not driven by the local optimality. It can be argued that similar to many other bio-inspired algorithms it is driven in the majority by the best results. However, the simple fact is that it is not directed toward the best solution but tries to make the best solution even better. GA mainly works out on the generation of a population pool by using mechanisms such as selection, use crossover [18–20], and mutation, and each of them can be regarded as random search mechanisms in the multidimensional search space. However, if we consider evolutionary strategies, the basic difference is that GA rely on the crossover [21], a probabilistic mechanism in which there is a useful exchange of information among solution sets and subsets to locate better solutions, whereas evolutionary strategies use mutation as the primary search mechanism. It must be mentioned here that in GA the parameters and their decision-making are very important than the operation(s) itself and hence the choice of the parameters and may be the involvement of more number of such parameters can enhance the performance of the GA. The parameters referred here are the internal parameters of the GA and must not be confused with the parameters of the problem. These parameters are the several decision-making parameters of the algorithm. They are the deciding factors for the combination, and thus the convergence and search mechanism for the problem. Hence, there is a requirement to discuss how the various parameters act and react for the solutions. The various parameters and operations of the GA cannot be separated, and hence as the terms arrive we will discuss their varieties. However, the main operators of the GA are as follows.

3.2 Encoding Schemes

Before understanding the operations, we must have a clear idea of the solution representations, and this will clarify how the different operations are taking place on them. Solution subset or rather the individual components of the solution does not exist individually but in a row in which the positional information denotes some specific significance for the problem or the system. In general, the operations may involve the whole solution set but can bring change on a subset. Several such changes in a subset can accumulate for a big change, but it all depends on the requirements. So, it is clear that GA involves a slow operation for local search, and the accumulation of such small equivalent steps can cover a significant portion of the search environment. However, we will see subsequently that the mutation factor can bring about a significant change. For any multiobjective problem, the solution is multidimensional that consist of several constrained variables or integer values (denoting some specific event) or a combination of both [22–24]. There are several ways of representing the solution chromosomes, and some of them are as follows.

3.2.1 Continuous Value Encoding

Continuous value encoding is the original value (length) representation kept for all variables. But for the operation to be successful and keeping the normal trend and the significance of the individual solution subset, all the values present must be of equal digits and there must be an equal number of digits before the decimal for all the solution set so that there is a fixed length chromosome for each of them. It is not necessary that all the individuals of a chromosome be of the same length but each chromosome must have the same length for each respective variable, i.e., same length for first variable, same for second variable and so on. This may require padding with zeros from both ends. The significance of this kind of fixed chromosome is that when operations such as crossover will occur, then the position of the individuals can be demarcated, and the decimal places will not change for an individual. Change of a decimal place can be a problem for a constrained variable.

3.2.2 Binary Encoding

In binary encoding representation, the whole numerical string is converted into a binary form consisting of only 0s and 1s. However, care is taken that each of the variable or individual member of the chromosome has equal bit length so that the individuality is maintained and later the bit length can be converted back into numerical values. A very simple example is provided in Figure 3.1.

Another reason of equal length chromosome and equal length bit representation will make the chromosome crossover fair and justified, and new offspring will have the same representation as their parents. Figure 3.2 will make the scenario clearer.

3.2.3 Integer Encoding

Integer encoding scheme makes the solution to be separated individually from the decimal part. This is done because floating values take longer time for computation and also during the operations of the GA, the decimal point can create problems in interpretation of the individuals. In addition, in many cases, the decimal portion is not so important with respect to the value of the fitness and can easily be considered without the decimal portion. An example will make the encoding scheme clearer (Figure 3.3).

Chromosome A	11001001001100100100
Chromosome B	10010011110010011111

FIGURE 3.1
An expression to show binary coding for GA.

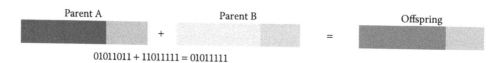

01011011 + 11011111 = 01011111

FIGURE 3.2
An example of genetic crossover for binary chromosomes.

Chromosome A	2 4 6 8 10 12 14 16 18 20
Chromosome B	6 7 2 4 9 5 1 3 8

FIGURE 3.3
An example of integer encoding.

3.2.4 Value Encoding or Real Encoding

In this kind of chromosome formation, the value of the solution individual consists of the real object representation or real values. The examples will make the concept clear. However, each of the chromosomes must be of equal length, and each position of the chromosome must denote some definite implications. In the first example, we have shown that each of the numbers must have equal length before and after the decimal place [25]. Even if that is not the case, positional information must be retained for each chromosome and must not be mixed up (Figure 3.4).

3.2.5 Tree Encoding

Tree coding is another type of representation, and the following example will make it clearer. The main aim of introducing different types of representations for chromosomes is to gather the convenience of each representation to represent a solution set for an application (Figure 3.5).

3.2.6 Permutation Encoding

Permutation encoding is denoted with the aim that the chromosome representation will have each of the integers from an initial to a maximum, and the range can be of any length. This is favorable for problems such as traveling salesman problem (TSP) and quadratic assignment problem (QAP) (Figure 3.6).

Chromosome A	11.365 56.937 28.984 39.347 23.554
Chromosome B	(up), (left), (up), (down), (right), (left)

FIGURE 3.4
Value encoding.

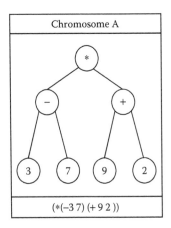

FIGURE 3.5
Tree encoding.

Chromosome A	5 10 6 9 3 7 2 8 11 4 12 1
Chromosome B	11 4 8 9 1 12 3 7 5 2 10

FIGURE 3.6
An example of chromosomes with permutation encoding.

3.3 Selection

Selection is an important operation for GA, and many researchers have found that the performance of GA depends highly on this. There are several selection procedures, and many of them are discussed here, but the point that must be considered is that each of them has their own limitations, and none of them is the sole winner. This is because of the probabilistic and the opportunistic model of the bio-inspired algorithm due to which it depends on the availability and involvement. This makes GA and all other bio-inspired algorithms probabilistic, and it is hard to predict the convergence, progress, stability, and success of the algorithm. However, in comparison with other deterministic, greedy, and dynamic programming models it can be said that one of the near best results is achieved in very less time.

During the crossover operation of GA, initially any two chromosomes must be selected from the chromosome pool and then the crossover feature gives rise to two new offspring from the existing ones. The two new offspring may or may not be better with respect to fitness, but the point is that the success of the two offspring depends on this selection feature. The more diverse is their selection; more diverse will be the offspring. In addition, a different combination of selections can give rise to the different quality of distribution of offspring, which will again help in the future generation selection and crossovers. So, what is required is such an algorithm for chromosome selection that is free from bias and provides an opportunity to each chromosome irrespective of their fitness. Several random selection-based algorithms are being proposed, and some of them are being discussed here with examples.

Before discussing how selection occurs, it must be noted that there are some calculations that are being made on each chromosome before any operation occur on them. Based on some numerical fitness evaluator, first the fitness values are calculated as $\{f_1, f_2, f_3, ..., f_n\}$ for n chromosomes and then probability of selection is calculated for each ith chromosomes as

$$(\text{Probability})_i = \frac{f_i}{\sum_{j=1}^{j=n} |f_j|} \qquad (3.1)$$

This equation assures that the best chromosomes will have a high probability of selection only when the problem is a maximization function. If the problem is a minimization problem, then we can take the inverse or place a minus sign before the fitness to calculate the probability of selection. The inverse will not cause a problem because many times it has been seen that even if the optimized value is zero, the solution never reached there and in that respect the inverse at any point of time will never create situations for infinity.

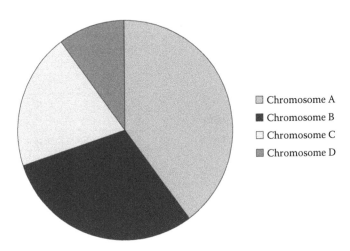

FIGURE 3.7
An example of roulette wheel selection.

3.3.1 Roulette Wheel Selection

The roulette wheel selection got its analogy from the wheel used in gambling. Roulette wheel selection is determined by a process in which the chance of selection of any one of the chromosomes for crossover is directly proportional to its fitness. This is a random selection like procedure in which the selection will sometime be reflected by the fitness of the chromosome and sometime may contradict it (Figure 3.7).

3.3.2 Rank Selection

Rank selection is a pretty straight selection of chromosomes in which the chromosomes are sorted and then the best lots are selected. The sorting is done based on the fitness evaluation of the chromosomes (Figure 3.8).

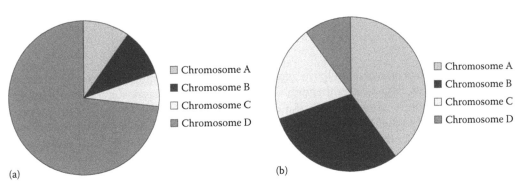

FIGURE 3.8
Rank-based selection: (a) Situation before ranking (group of fitness) and (b) situation after ranking (graph of order numbers).

— Given the following population and fitness

— Select C and E	Individual	Fitness	Probability
— E wins	A	5	1/25 = 4%
— Probability	B	6	3/25 = 12%
(2s − 2r + 1)/(s * s)	C	8	5/25 = 20%
where s = size of population	D	11	7/25 = 28%
r = rank of "winning	E	20	9/25 = 36%
individual"			

FIGURE 3.9
Example of tournament selection.

3.3.3 Tournament Selection

In the tournament selection, every two chromosomes are made to play a virtual confrontation with each other, and the winner gets through the selection process. An excellent example of such selection is making a toss of the two random generators that generate real values within the same range and the greater one wins (Figure 3.9).

3.3.4 Steady-State Selection

The main idea of this selection is that a large part of chromosomes should survive till the next generation. GA then works in a following way. In every generation, a few (good—with high fitness) chromosomes are selected for creating a new offspring. Then some (bad—with low fitness) chromosomes are removed and the new offspring is placed in their place. The rest of the population survives till the new generation.

3.3.5 Random Selection

Random selection involves the selection of the parents in a random manner and in this way each of the chromosomes is equally likely to be selected.

3.4 Crossover

The phenomenon of crossover helps in generating two new chromosomes from the existing two chromosomes. This occurs through some points in the chromosome body. However, that point position is same for both the chromosomes otherwise the offspring chromosome will be of unequal length. Unequal length in chromosome will result in a distorted solution. Figures 3.10 through 3.15 and the example will provide a better view and perception of the crossover phenomenon.

3.4.1 Single Point Crossover

In a single point crossover, the crossover takes place at only one point. The following example will make it clearer.

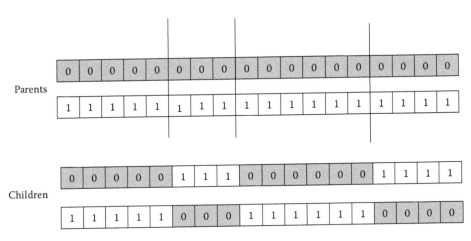

FIGURE 3.10
An example of N point crossover.

3.4.2 N Point Crossover

Crossover can take place from any point, and there can be a multiple number of points. However, the number of points can vary but must be less than the number of points that are available in the chromosome (Figure 3.10).

3.4.3 Uniform Crossover

The uniform crossover uses a fixed mixing ratio between the two parents. Unlike one- and two-point crossover, the uniform crossover enables the parent chromosomes to contribute the gene level rather than the segment level (Figure 3.11).

3.4.4 Arithmetic Crossover

In an arithmetic crossover, the physical crossover phenomenon is being replaced by a mathematical operations such as the OR, AND, or XOR, or some complex mathematical equations. The operation will depend on the physical representation of the chromosome and on what kind of operation or mathematical modification is feasible and permissible for the chromosome (Figure 3.12).

$$11001011 + 11011101 = 11011111 \text{ (OR)}$$

FIGURE 3.11
An example of uniform crossover.

$$11001011 + 11011111 = 11001011 \text{ (AND)}$$

FIGURE 3.12
An example of arithmetic crossover.

3.4.5 Tree Crossover

The tree crossover is another kind of crossover in which the tree structures (representing the chromosomes) get crossed giving rise to two new chromosomes. However, as the structure of the tree crossover is nonlinear there is a high probability that the newly generated chromosomes may become meaningless or devoid of any concrete meaning in regard to the solutions for the application (Figure 3.13).

3.4.6 Order Changing Crossover

In an order changing crossover, the resultant offspring is generated through the changing of the order of the solution chromosome as a whole or a subset of the solution chromosome (Figure 3.14).

3.4.7 Shuffle Crossover

In a shuffle crossover, there will occur a shuffling of the position of the variables and this will give rise to a new chromosome. However, if the range or the constraints on each variable is different, shuffle crossover can lead to an invalid or unfeasible chromosome.

3.5 Mutation

Mutation introduces diversity into the population. However, it has its own limitations. Sometimes it is said that the mutation operator helps in escaping from the local minima to the global minima. There are various types of mutation operator and is very specific to the applications sometimes. However, it must be said that there are various ways of perceiving the conception and get what is required for the chromosomes of the problem.

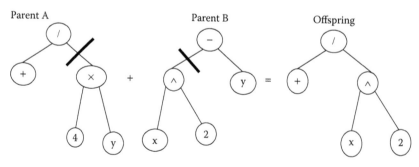

FIGURE 3.13
An example of tree crossover.

$$(1\ 2\ 3\ 4\ 5\ 6\ 8\ 9\ 7) \quad \Rightarrow \quad (1\ 8\ 9\ 4\ 5\ 6\ 2\ 3\ 7)$$

FIGURE 3.14
An example of order changing crossover.

The following is a list of mutation operators that are used for permutation representations:

- Inversion mutation
- Insertion mutation
- Displacement mutation
- Reciprocal exchange mutation (swap mutation)
- Shift mutation

3.5.1 Inversion Mutation

The inversion mutation operator randomly selects two points in the chromosome. The allele values that are in between these two points are then inverted to form a new chromosome. Figure 3.15 shows an example of this mutation.

3.5.2 Insertion Mutation

The insertion mutation operator selects a gene at random and then inserts it at a random position.

3.5.3 Displacement Mutation

The displacement mutation operator randomly selects two points in a chromosome and then moves the substring to another randomly selected location in the chromosome. This mutation operator is similar to the insertion operator, but instead of moving only one allele, a substring is moved. Figure 3.16 shows an example of the displacement mutation operator.

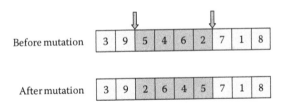

FIGURE 3.15
Example of inversion mutation.

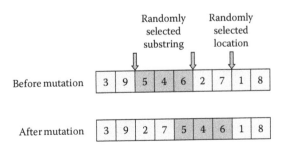

FIGURE 3.16
Example of displacement mutation.

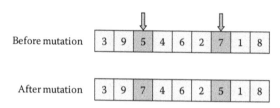

FIGURE 3.17
Example of reciprocal exchange mutation.

3.5.4 Reciprocal Exchange Mutation (Swap Mutation)

The reciprocal exchange mutation operator (also known as swap mutation) selects two alleles at random and then swaps them. Figure 3.17 shows an example of this type of mutation.

3.5.5 Shift Mutation

The shift mutation operator selects a gene at random and then shifts the gene at a random number of positions to the right or left of its present spot. Figure 3.18 shows an example of this mutation.

3.6 Similarity Template

The schema can be defined as a similarity template in which subsets of the chromosome have similarities in terms of elements at certain similar positions. This similarity template of a schema represents a subset of all chromosomes that have the same bits at certain positions and is important because we can associate the fitness value of the chromosome with the schema's defining length and associate an average fitness of the schema. For a given population, this is important as this average fitness of the instances of the schema varies with the population and we can track the composition of the generation and its rate of change of the quality of the population.

A schema represents a general class of chromosomes. A binary schema consists of a string over the alphabet {0,1,*}; where "*" represents "don't care" bits and could be either "0" or "1." Multiple strings in the search space belong to one schema. Positions in the string in which "0" or "1" occurs are termed as fixed positions. Schema theory deals with the study of the existence of a schema in the population at various generations and with

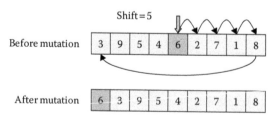

FIGURE 3.18
Example of shift mutation.

the analysis of the effect of selection, crossover, and mutation on the chance of survival of a schema. The number of fixed positions of a schema is its order. However, a schema's defining length is the distance between the outermost fixed positions. Start searching from the left for the first fixed position and then reach the least one and the difference between the positions of the extreme right and the extreme length provides us the defining length. The higher value for defining length denotes that the chromosome must undergo more on mutation than on crossover. Some of the key terms related to schema are defined below:

Definition 1:   Order of a Schema

The order of a schema $o(H)$ is the number of fixed positions in the string.

Definition 2:   Defining Length of a Schema

The defining length of a schema $d(H)$ is the distance between the first and the last fixed positions in the string, that is, the number of places where crossover can disrupt the schema.

Definition 3:   Fragility of a Schema

The fragility of a schema is the number $d(H)/(l-1)$, (where, $d(H)$ is the defining length and l is the length of the schema), that is, the proportion of places where the schema can be disrupted.

Example

Suppose, we have a schema $H_1 = 10**0 * 1***$.
For this schema,
Length $l(H_1) = 10$ (total number of bits in the string representing it)
Order of schema $o(H_1) = 4$ (Number of fixed bits)
Defining length $d(H_1) = 7 - 1 = 6$ (Difference between the last fixed bit and the first fixed bit)
Vulnerability or fragility $d(H_1)/(l-1) = 6/9 = 0.66667$
Vulnerability represents the portion of the schema in which the crossover will destroy it.
Some of the possible instances of schema H_1 are represented by the following set:
I = {1000011110, 1011011111, 1010001001, 1011011000}.

3.7 Building Blocks

In GA crossover tends to conserve the genetic information present in the strings to be crossed. Thus, when the strings to be crossed are similar, its capacity to generate new building blocks diminishes. Mutation is not a conservative operator and can generate radically new building blocks. Selection provides the favorable bias toward building blocks with higher fitness values and ensures that they increase in representation from generation-to-generation. GA's crucial and unique operation is the juxtaposition of building blocks achieved during crossover and this is the cornerstone of GA mechanics.

3.8 Control Parameters

Following are the control parameters of GA:

- *Population diversity*: This controls the amount of variation that is allowed among the solutions. If the average distance between individuals (also known as variation) is large, the diversity is large. If, on the other hand, the average distance is small, the diversity is low. Getting the right amount of diversity is an essential prerequisite to effective solutions because if the diversity is too high or too low, the GA will not work well.

- *Crossover probability*: This controls the fraction of children that are crossover children. A crossover fraction of 1 means that all children other than elite individuals are crossover children, whereas a crossover fraction of 0 indicates that all children are mutation children.

- *Mutation parameters*: There are two parameters that control the amount of mutation allowed:
  - *Scale*: Controls the standard deviation of the mutation at the first generation.
  - *Shrink*: Controls the rate at which the average amount of mutation decreases.

- *Fitness scaling*: Fitness scaling converts the raw fitness scores that are returned by the fitness function to values in a range that is an amenable selection function.

- *Mutation probability*: Says how often a chromosome will be mutated. If there is no mutation, offspring is taken after crossover (or copy) without any change. If mutation is performed, part of a chromosome is changed. If mutation probability is 100%, whole chromosome is changed, if it is 0%, nothing is changed at all.

- *Population size*: It describes how many chromosomes are in a population (in one generation). If there are too few chromosomes, GA becomes slow and vice versa.

- *Number of generations*: The number of generations is closely related to improvement in the fitness function. A fitness function usually shows a major improvement in early generations and then asymptotically approaches an optimum.

3.9 Nontraditional Techniques in GAs

Each of the chromosomes in GA is a set of solution individuals where if S is the solution set then $S = \{s_1, s_2, .., s_k\}$, where s_k is the kth solution individual, and k is the length of the string or the maximum number of possible events without repetition. Throughout the book, the solution considered is such a set and the terms *Solution, Solution String,* and *String* are used interchangeably to denote the same meaning. The length of the solution string depends on the problem or the dimension of the considered equation. In many problems, it can be of variable length and depends on the dynamic development and improvement of the solution string. However, each position in the string has two meanings. For an equation-based problem the position denotes a certain variable, whereas for graph-based and combinatorial optimization problem the sequence is more important than the positions. That means the position is relative to the position of other solution individuals.

Steps for the Traditional Generalized Genetic Algorithms

1. Procedure{GA}{graph, agents, globaloutput}
2. Conversion: Problem Model as a Genetic Algorithms Problem
3. Initialization Chromosomes, Solution Variables
4. Iteration1 = 1 to N1
5. Iteration2 = 1 to N2
6. Genetic Algorithms Selection Procedure
7. Apply Genetic Algorithms Crossover Operator
8. Apply Genetic Algorithms Mutation Operator
9. EndFor
10. Evaluate Each Individual's Fitness of New Chromosomes
11. Update the Best Chromosome
12. Form New population for next Iteration1 Generation
13. EndFor
14. The Best Chromosome is the Solution
15. EndProcedure

3.9.1 Genetic Programming

GP involves creating programs for the given task automatically. Gandomi and Alavi proposed a multistage genetic programming as an improvement to genetic programming (GP) for a nonlinear system modeling. It is based on incorporating the sole effect of predictor variable as well as the interactions between the variables to provide more accurate simulations. The mentioned algorithms have been widely used to solve a different kind of optimization tasks.

3.9.2 Discrete Genetic Algorithms

GA is ideal for the discrete problems and can easily be used in various graph-based problems and steady-state space-based problems. The main reason for utilization of the GA for discrete problems is the absence of any variation-based operators. The advantages of GA are the operators, which can create a better combination of the solution subset. Like for binary chromosome-based GA we can easily use for the 0/1 knapsack problem or similar problems in which the binary sequence can make sense, and the applicability of the GA will go up only when the problem and its solution formation can be replicated into such a structure in which GA can be easily utilized and most important the sequence of the solution is actually making sense and is not violating the constraints of the problem.

3.9.3 Genetic Algorithms for Continuous Applications

GA is highly applicable for the continuous application so far as the high-dimensional problems are concerned. But compared to other algorithms such as particle swarm optimization (PSO) algorithm and similar swarm intelligence algorithms GA is relatively less capable of variation of the individual solution subset. However, the combination generation capability of the GA is highly respectful and elaborative.

3.10 Convergence Analysis of Genetic Algorithms

Convergence of the GA for problems is highly dependent on the combination generation capability, which is again dependent on the opportunity, which has arisen for the operators such as crossover and mutation. There are various types of such operator, and each of them uses a certain technique for decision-making of how the operations must take place. Each of such type ensures that the combinations generated in each of the iteration are unique and different from the others. This must accompany randomness in selection but yet must be deterministic so that the desirables chromosomes get their opportunity to participate more than the undesirable ones. This will guarantee that there is a high probability of better and better solution in the future generations. The criteria for such enhancement and which decision criteria must be chosen are subjected to experimentation or in some cases and experimentation it has been seen that the incorporation of all the criteria can generate a better result. The trade-off is made by the increase in complexity of the computation. These combination generation decisions and success and the continuously enhanced solution generation will decide the convergence of the GA and highly depend on the dataset and mathematical formulation of the application.

3.11 Limitations and Drawbacks of Genetic Algorithms

Some of the limitations of the GA are as follows:

- First and the foremost drawback is the lack of the variation parameters for the individual solution chromosomes.
- The initial solution chromosomes are vital so far as the convergence and next-generation chromosomes are concerned.
- The mutation parameters must be specific and predetermined else it can worsen the solution.
- The crossover rate is very slow for the convergence and local search. Deterministic approach fails for the operations.
- The bound or the constraints for the algorithms are difficult to handle or rather it can be said that each must be taken care of individually.

3.12 Summary

The following concepts have been covered in this chapter:

- The fundamentals of GA and how a problem can be represented and solved using it.
- The various methods through which we can represent a chromosome, such as real coded chromosome, binary coded chromosome, continuous value encoding, binary encoding, integer encoding, value encoding, tree encoding, and permutation encoding.
- Different selection schemes such as roulette wheel selection, rank selection, tournament selection, steady-state selection, random selection, and fitness-based selection.
- Various types of crossovers such as single point crossover, N-point crossover, uniform crossover, arithmetic crossover, tree crossover, bit inversion crossover, order changing crossover, segmented crossover, and shuffle crossover.
- Types of mutations such as bitwise inversion, string wise inversion, and random selection-based inversion.
- Concepts related to several issues such as similarity template, competition, building blocks, schema theorem, control parameters, and nontraditional techniques in GA.
- GA for discrete as well as continuous applications.
- Convergence analysis of GA and how it influences the optimization process.
- Limitations and drawbacks of GA for specific problems and representations.
- GA with uneven chromosome length for certain special applications such as graph-based problems.

Solved Examples

1. How can we optimize the features in gait database using GA algorithm? Illustrate the same using benchmark datasets.

Solution

GA optimization technique is applied on the datasets of Parkinson patients. The purpose of this algorithm is to either optimize or reduce the features required for the recognition of Parkinson disease. As the class labels to which a dataset entry belongs to is known a priori, we appended limited number of bits for assignment of class labels. To each instance that belongs to the patient affected with Parkinson disease, a binary "1" is appended else a binary "0" is appended.

The experiments can be carried out in any compatible IDE like Eclipse and so on. The first dataset that we consider is of PHYSIONET gait dataset. GA algorithm is applied over the dataset and only relevant features are extracted. The selected feature subset (as only relevant features have been extracted) obtained from GA algorithm is applied. The results are summarized in Table 3.1 and Figures 3.19 through 3.21.

TABLE 3.1

Classification Accuracy of GA Algorithm over Optimized
Physionet Dataset (Dataset-I)

Algorithm	Dataset Name	Accuracy (%)
GA	Ga	83.90
	Ju	93.72
	Si	89.93

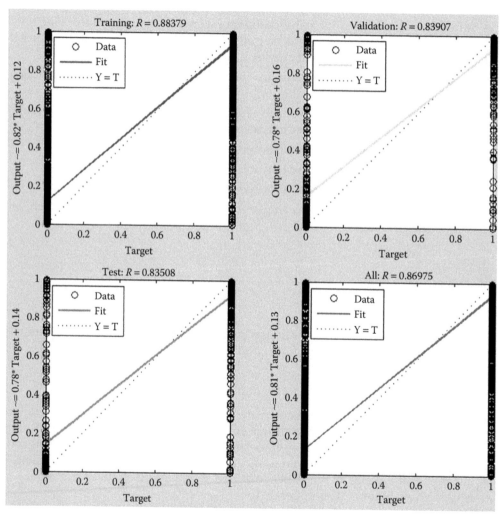

FIGURE 3.19
Regression plot of Ga dataset (GA Algorithm).

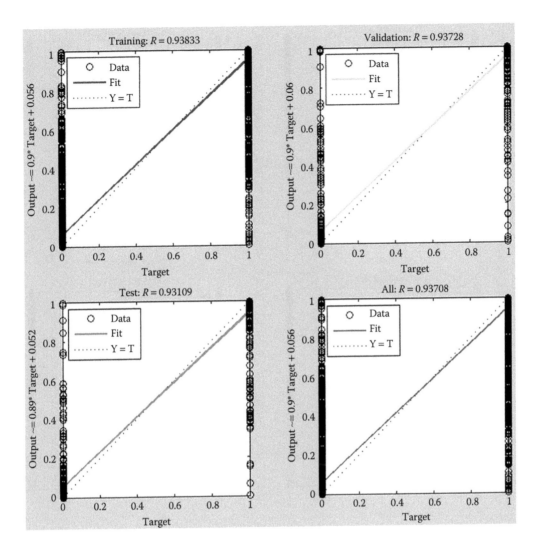

FIGURE 3.20
Regression plot of Ju dataset (GA Algorithm).

A similar approach was also applied on a second benchmark dataset, that is, UCI dataset. The results could be summarized in Table 3.2 and Figure 3.22.

2. Discuss the role of crossovers in GA. How are double point crossovers better than single point crossovers?

If crossover is excluded from GA, they become something between the gradient descent and the simulated annealing. The main effect of crossover consists of the exchange of parts of different solutions. If an optimization task can be loosely decomposed into somewhat independent subtasks, and this decomposition is reflected in genes, then crossover will increase the performance of GAs. For

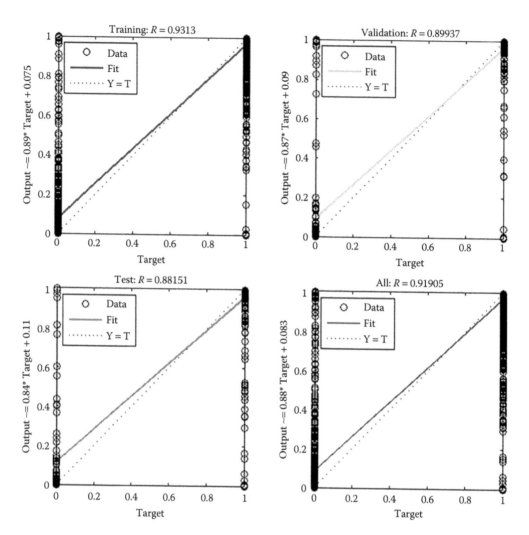

FIGURE 3.21
Regression plot of Si dataset (GA Algorithm).

example, if there is a function $f(x,y) = g(x) + h(y)$, and x and y are encoded consequently in the genome, and for example, $g(x)$ has larger influence, then the part of
genome that stands for x will be optimized in the first place, and it will become
nearly the same for the whole population; thanks to crossover. After this, $h(y)$ term
will be optimized. That is, crossover helps to loosely decompose the task into subtasks without prior knowledge (but dependent on the encoding scheme), if it is
possible (otherwise it will yield no benefit, or even will make search less efficient).
This is actually the main additional meta-heuristic of GAs in comparison with
other meta-heuristic optimization methods.

Double-point crossover (or other more *clutter* types of crossover) can be more
beneficial, if the most crucial parts of solutions are encoded in the middle of the
genome or if they are placed in separate locations. The simplest example is the
function $f(x,y,z) = h(x,z) + r(y) + g(z)$, where $h(x,z)$ has the highest influence, and

TABLE 3.2

Classification Accuracy of GA Algorithm over Optimized
UCI Dataset (Dataset-II)

Algorithm	Dataset Name	Accuracy (%)
GA	UCI	79.93

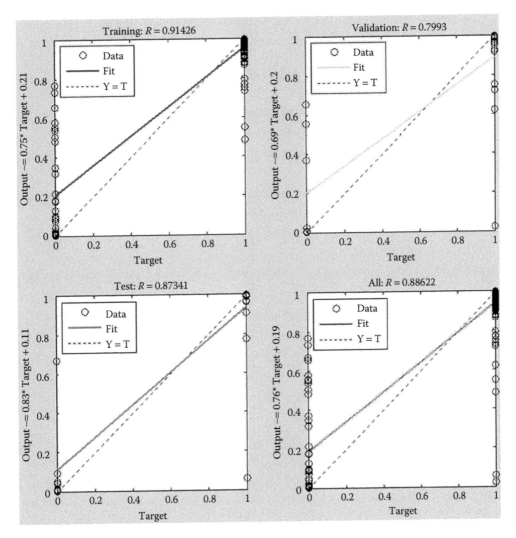

FIGURE 3.22
Regression plot of UCI dataset (GA Algorithm).

genes encode the string *xyz*. Single-point crossover will prefer to take *x* from one
parent, and *z* from another parent. However, (*x,z*) determine the solution quality
together. If one individual will occasionally have optimal components (*x*,z**), they
will not appear simultaneously in its children after single-point crossover, so it
will be more difficult for these parts of the genome to stabilize.

Exercises

1. What is GA? Explain its resemblance with natural evolution.
2. What are the basic operators used in GA? Explain the importance of each one of them.
3. Discuss the various selection schemes for GA.
4. Explain the concept of roulette wheel selection with the help of a suitable example.
5. What are the various types of crossover operations? Explain the working of each one of them.
6. Illustrate the working of a two-point crossover operator with the help of suitable examples.
7. How is diversity introduced into the population in GA?
8. What is the role of mutation in GA?
9. Describe the different types of mutation operators with the help of suitable examples.
10. What is a schema?
11. Derive the expression for schema survival after traditional crossover and mutation.
12. Discuss the various application domains of GA.
13. How is a GA different from traditional search techniques?
14. What is a chromosome?
15. What do you mean by encoding? Discuss binary encoding with suitable examples.
16. What is a gene?
17. Explain the stochastic nature of GA.
18. Despite their stochastic nature, why are GA successful?
19. Which operator in GA is basically responsible for exploration? How?
20. What will happen if the *mutation rate* will be too high? Justify your answer.

References

1. S. García, D. Molina, M. Lozano and F. Herrera, A study on the use of non-parametric tests for analyzing the evolutionary algorithms' behaviour: A case study on the CEC'2005 special session on real parameter optimization, *Journal of Heuristics*, 2009, 15(6), 617–644.
2. F. Herrera, M. Lozano and J. L. Verdegay, Tuning fuzzy logic controllers by genetic algorithms, *International Journal of Approximate Reasoning*, 1995, 12, 299–315.
3. A. Mishra and A. Shukla, Mathematical analysis of the cumulative effect of novel ternary crossover operator and mutation on probability of survival of a schema. *Theoretical Computer Science*, 2016, 1–11. doi:10.1016/j.tcs.2016.07.035.
4. A. Mishra and A. Shukla, Mathematical analysis of schema survival for genetic algorithms having dual mutation. *Soft Computing*, 2017, 1–9. doi:10.1007/s00500-017-2650-3.
5. J. R. Cano, F. Herrera and M. Lozano, Using evolutionary algorithms as instance selection for data reduction in KDD: An experimental study, *IEEE Transactions on Evolutionary Computation*, 2003, 7, 561–575.

6. F. Herrera, M. Lozano and J. L. Verdegay, A learning process for fuzzy control rules using genetic algorithms, *Fuzzy Sets and Systems*, 1998, 100, 143–158.

7. A. Mishra and A. Shukla, Analysis of the effect of elite count on the behavior of genetic algorithms: A perspective. *7th IEEE International Advance Computing Conference (IACC-2017)*, VNR Vignana Jyothi Institute of Engineering and Technology, Hyderabad, India, January 2017. doi:10.1109/IACC.2017.0172.

8. S. Maheshwari and A. Tiwari, A novel genetic based framework for the detection and destabilization of influencing nodes in terrorist network. *In Computational Intelligence in Data Mining*, Volume 1, 2015, pp. 573–582. Springer, New Delhi, India.

9. M. Lozano and C. G. Martínez, Hybrid metaheuristics with evolutionary algorithms specializing in intensification and diversification: Overview and progress report, *Computers & Operations Research*, 2010, 37, 481–497.

10. C. M. Fonseca and P. J Fleming, Genetic Algorithms for Multi-Objective Optimization: Formulation, Discussion and Generalization, In *Proceedings of the 5th International Conference on Genetic Algorithms*, pp. 416–423. Morgan Kaufmann Publishers, San Francisco, CA, 1993.

11. A. Konak, D. W. Coit and A. E. Smith, Multi-objective optimization using genetic algorithms: A tutorial, *Reliability Engineering & System Safety*, 2006, 91, 992–1007.

12. P. Moscato, On evolution, search, optimization, genetic algorithms and martial arts. Technical Report C3P 826, Caltech Con-Current Computation Program 158-79. California Institute of Technology, Pasadena, CA, 1989.

13. L. Davis, *Genetic Algorithms and Simulated Annealing*, Pitman, London, UK, 1987.

14. J. J. Grefenstette, Optimization of control parameters for genetic algorithms, *IEEE Transactions on Systems, Man, And Cybernetics*, 1986, 16, 122–128.

15. N. Srinivas and K. Deb, Multi-objective optimization using non-dominated sorting in genetic algorithms, *Journal of Evolutionary Computation*, 1994, 2, 221–248.

16. C. G. Martínez, M. Lozano, F. Herrera, D. Molina and A. M. Sánchez, Global and local real-coded genetic algorithms based on parent-centric crossover operators, *European Journal of Operational Research*, 2008, 185, 1088–1113.

17. F. Herrera, M. Lozano and A. M. Sánchez, Hybrid crossover operators for real-coded genetic algorithms: An experimental study, *Soft Computing*, 2005, 9, 280–298.

18. C. K. Peng, S. Havlin, H. E. Stanley and A. L. Goldberger, Quantification of scaling exponents and crossover phenomena in non-stationary heartbeat time series, *Chaos: An Interdisciplinary Journal of Nonlinear Science*, 1995, 5, 82–87.

19. I. M. Oliver, D. J. Smith and J. R. C. Holland, Study of permutation crossover operators on the travelling salesman problem, *Proceedings of the Second International Conference on Genetic Algorithms on Genetic Algorithms and their Application*, July 28–31, 1987 at the Massachusetts Institute of Technology, Cambridge, MA 1987. Hillsdale, NJ: L. Erlhaum Associates, 1987.

20. W. Chang, A multi-crossover genetic approach to multivariable PID controllers tuning, *Expert Systems with Applications*, 2007, 33, 620–626.

21. F. Herrera, M. Lozano and A. M. Sánchez, A taxonomy for the crossover operator for real-coded genetic algorithms: An experimental study, *International Journal of Intelligent Systems*, 2003, 18, 309–338.

22. M. Lozano, F. Herrera, N. Krasnogor and D. Molina, Real-coded memetic algorithms with crossover hill-climbing, *Evolutionary Computation*, 2004, 12, 273–302.

23. F. Herrera and M. Lozano, Gradual distributed real-coded genetic algorithms, *IEEE Transactions on Evolutionary Computation*, 2000, 4, 43–63.

24. M. Lozano, F. Herrera and J. R. Cano, Replacement strategies to preserve useful diversity in steady-state genetic algorithms, *Information Sciences*, 2008, 178, 4421–4433.

25. F. Herrera, M. Lozano and J. L. Verdegay, Tackling real-coded genetic algorithms: Operators and tools for behavioural analysis, *Artificial Intelligence Review*, 1998, 12, 265–319.

4

Ant Colony Optimization

4.1 Introduction

Ant colony optimization (ACO) is an agent-based cooperative search meta-heuristic algorithm, in which each agent seeks its own independent solution under the influence of pseudo communication with social or global fellow agents [1]. ACO was initially proposed by Marco Dorigo in 1992 in his PhD thesis as an ant system, which imitates the habit of ants in a food search-cum-procurement and helping the fellow ants to the food source [2]. Till then, it has emerged as one of the best algorithms for optimization mainly for the graph-based discrete problems.

There are a few types of variants of ACO, which are prevalent and have unique mathematical features to enhance the performance of the algorithm and at the same time balance between exploration and exploitation. ACO is such an algorithm which is perhaps suitable for any kind of application [3–6]. However, there are some drawbacks in the algorithm, which have been overcome by hybridization with other nature-inspired algorithms. ACO does not possess any operator, which can vary a variable parameter through a continuous space. This makes the algorithm cripple at situations in which there is requirement of continuous search. Another drawback is encountered during the path planning or adaptive sequence generation in which the solution is the sequence of events (or states or nodes), the number of which is not fixed. If ACO is implemented for such kind of applications then there is a possibility that proper sequence as valid solution can be created, but chances are there that a loop is formed. In such case, the agent is wasted.

The main features of the ACO are the simple mathematical model and its associated dependency generation, which other agents can thrive on. For this, ACO uses another set of memory associated with the locations of the workspace for pheromone-level registration and monitoring function, so as to update the value, which decreases with time following a mathematical integrity and not randomly but increases when any ant releases some pheromones following another mathematical relation at that location of search space.

4.2 Biological Inspiration

It is a well-known fact that ants, wasps, and so on are known to show some of the most complex process ever found in the insect world [7]. Their behavior is dictated by the collectiveness of their communities to achieve as well as sustain vital tasks as defense, feeding, reproduction, and so on. Though complete understanding of ant colonies is near to impossible, yet there are some aspects that are well known to the researchers and that have proven to be consistent. These aspects include the following:

4.2.1 Competition

Ants live in small to large-sized communities within the environment. This environment in which they live affects their daily functioning and the tasks they perform.

4.2.2 High Availability

Gordon has conducted a lot of experiments to study the behavior of the ants, and his experiments have shown that ants may also switch tasks [7] (It is basically expected that the ants will keep the same task attributions). There are four different tasks that an ant can perform:

1. Foraging
2. Patrolling
3. Maintenance work
4. Middle work

Several studies have proven that in certain species, a majority of ants remain inside the nest just sitting idle with no task assigned to them [8]. For instance, in a 20,000 ant's colony, only 5,000 ants perform external work, whereas the other 15,000 ants remain idle within the nest. A theory suggests that these ants remain idle so that they can contribute to the colony when there is some contingency such as a flood and so on.

4.2.3 Brownian Motion

In their natural habitat, ants are continuously inundated with stimuli of all kinds. These stimuli are presented to them from all around, and a great part of their sensorial system is dedicated to pick up only the most important stimuli [9]. These stimuli include heat, magnetism, light, pressure, gravity, and chemicals. Recently, chemicals have become one of the major stimuli because of the effects that it accompanies with itself. It can be safely asserted that under the influence of any stimuli (except gravity), ants tend to show a random pattern of locomotion. If a stimulus provides a particular orientation to this locomotion, then its absence does not remove such pattern; instead, it provides another random pattern of locomotion. A number of experiments have been performed to observe this phenomenon.

The following figure demonstrates a pattern followed by the ants when they perform a Brownian motion (Figure 4.1).

Thus, in essence, ants remain very determined when they are able to sense a pheromone or food around. Even when a pheromone trail is obstructed, they will still find the way to continue their trail. It is also to be noted that in the real world, the sequence of movements performed depends on the task at hand. For example, when the ants are patrolling, they

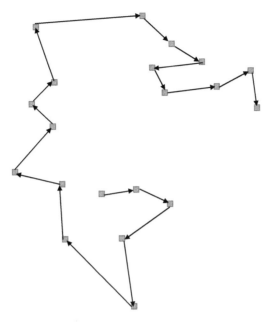

FIGURE 4.1
Brownian motion shown by the ants.

show a locomotion pattern, which is similar to *Brownian movement*. Similarly, maintenance and middle workers usually do not go very far from the nest, whereas the foragers leave the nest usually in a straight line.

4.2.4 Pheromones and Foraging

Pheromone is a generic name and refers to any endogenous chemical substance that is secreted by an insect. Insects secrete this chemical in order to incite reaction in other organisms of the same species. Insects use pheromones for diverse tasks that include reproduction, identification, navigation, and aggregation [10]. The antenna found on the ant body is the agent responsible for detecting pheromones. The chemoreceptors present on the antenna trigger a biological reaction when the sum total of all action potentials exceeds a given threshold.

Different species of ants have different strategies for foraging [1]. Most species have some selected ants who have been assigned the task of scanning the environment around for food. When they leave their nest, they usually take one direction and explore the land in this direction. A pheromone trail is left behind them and this trail is used by other ants if the foraging ant happens to succeed in finding the food. It is a common observation that the foragers that select shorter paths return to their nest earlier; as a result, they deposit more pheromone into that trail. Ants leaving the nest, next, choose the trail probabilistically (the one with a higher concentration of pheromones is chosen).

4.3 Basic Process and Flowchart

Steps for the Traditional ACO Algorithm Can Be Summarized as Follows:

Procedure{Traditional ACO} (graph, agents, globalOutput)
Conversion: Problem Model as a Graph Problem G (V, E)
Initialization Graph Matrix, Ant-Agents, EdgePhLevel
For (Ant-Agents = 1 to N)
       Calculate Probability of Movement for Each Edge or Subset of Edges
       Move To the new node
       Update the Memory for Path Traversed
       Update Pheromone Level for Edges
EndFor
For
       If (Destination Reached or Solution Complete)
          Calculate fitness
          If (Fitness Is Better Than Global Best)
             Update Global Best
          EndIf
       EndIf
EndFor
EndProcedure

Alternatively, the following flowchart highlights the steps involved in the traditional ACO algorithm (Figure 4.2) [11].

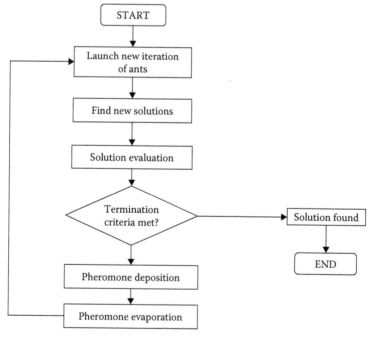

FIGURE 4.2
Flowchart of ant colony optimization algorithm.

4.4 Variants of Ant Colony Optimization

ACO has quite a number of variants, which have some modified pheromone updation rule, which enhances the scope of more exploration than exploitation and many researchers claim that there must be a balance between the two and the mode of considerable exploration will create an opportunity for better global optimization and the fellow agents will be less bias from the already created path to the goal.

4.4.1 Ant System

Marco Dorigo took the leading role in introducing the ant system (AS) meta-heuristics. Basically, the initial algorithm consisted of n number of ants, which move on a certain environment on the basis of pheromone trail updated through τ_{ij} and the local heuristic information given by η_{ij}. This approach helped in solving combinatorial optimization problems, which cannot be solved by simple traditional algorithms. The pheromone updation scheme is controlled by all the ants taking part in the search process.

4.4.2 Ant Colony Optimization

ACO was the next development of the ant system (AS). There is however some controversy on the meaning of the term *Ant Colony Optimization*. The term is sometimes used as a generalized term for all the AS algorithms and its modified versions as a whole. However, some researchers claim that in ACO, only the ant having found the best solution is exploited and is allowed to update the pheromone trails pretty much similar to the modified algorithms described in the subsequent paragraphs.

The density of pheromone secretion in a AS indicates intensity of favorable conditions and opportunity of the aim getting fulfilled and hence increases or decreases the chance or probability of other ants following the same path. Secretion of pheromone of the ants is spontaneous and depends on the density of the ants and the time factor, that is, how long the ant was present on that path and is averaged and scattered throughout the path, instead of introducing some exponential function. This leads to the conclusion of shortest path being planned, as it will lead to a more number of ants passing through a point of the shortest path.

The traditional equation of pheromone updation is

$$\tau_{xy}(t+t') \leftarrow \begin{cases} (1-\rho t')\tau_{xy}(t) + \sum \Delta\tau_{xy}^k(t) & \text{if ant passes} \\ (1-\rho t')\tau_{xy}(t) & \text{if no ant passes} \end{cases} \tag{4.1}$$

where $\tau_{xy}(t)$ is the pheromone deposit concentration level by all k ants at any path between node x and y, $\Delta\tau_{xy}^k(t)$ is the pheromone deposited by the kth ant in path between node x and y, $0 < \rho < 1$ is the evaporation rate and is different for different breed, t is the time factor, and t' is the estimated time interval.

The pheromone deposited by the kth ant in the path between node x and y is given by

$$\Delta\tau_{xy}^k(t) = \begin{cases} \dfrac{Q}{L_k(t)} & \text{if } k\text{th ant passes at time } t \text{ through path } xy \\ 0 & \text{otherwise} \end{cases} \tag{4.2}$$

where Q is just a constant to enhance the pheromone deposit and depends on the modeling, rate of evaporation, and also the system and its associated environment. $L_k(t)$ factor is the optimization function or some function associated with it. The optimization function must be a minimum best optimization function, for the mathematical model to work. However, if there is a maximum best optimization function, then the equation will be the following:

$$\Delta \tau_{xy}^k(t) = \begin{cases} Q * L_k(t) & \text{if } k\text{th ant passes at time } t \text{ through path } xy \\ 0 & \text{otherwise} \end{cases} \tag{4.3}$$

Now, there are uncertainties in finding the optimized solution, as no matter how many ants are on the mode of exploration and exploitation, situation may arise when the ant agents may get deviated from the right path and also the dynamic parameters can change, and thus the cost function may evaluate the planned path as nonoptimized.

To handle this kind of situation, the simulation is carried out considering that the parameters can remain constant for at least a certain time interval for evaluation purpose and efficiency consideration.

4.4.3 Best–Worst Ant System

The features of the best–worst ant system (BWAS) are described. Here, the pheromone concentration update rule goes as follows:

$$\tau_{ij}(t+1) = \tau_{ij}(t) + \Delta \tau_{ij} \tag{4.4}$$

where

$$\Delta \tau_{ij} = \begin{cases} \Phi\{\Psi(\text{ant}_{\text{Global–Best}})\} & \text{for } (i,j) \in \text{ant}_{\text{Global–Best}} \\ 0 & \text{otherwise} \end{cases} \tag{4.5}$$

In addition, $\Phi\{\}$ is the pheromone deposit function, which depends on the evaluation function $\Psi\{\}$ and the best global solution presenting ant for the path ij.

Another important component of the BWAS is the pheromone trail mutation, in which the pheromone concentration suffers mutation or a sudden change in the concentration occurs probabilistically to introduce a new diversity in the search algorithm.

The mathematical model for mutation, which operates with a probability p_m is as follows:

$$\tau_{ij}^m = \begin{cases} \tau_{ij} + \text{Mutation}\{i, \tau_{\text{treshold}}\}, \text{if } a = 0 \\ \tau_{ij} + \text{Mutation}\{i, \tau_{\text{treshold}}\}, \text{if } a \neq 0 \end{cases} \tag{4.6}$$

where τ_{ij}^m is the mutated value for τ_{ij}, mutation$\{\}$ is a function of the current iteration i, τ_{treshold} is the average pheromone deposit trail, and a is a randomly generated measure of chance of occurrence of mutation.

Pheromone Trail Mutation: The pheromone trail suffers from unexpected mutations [12]; these mutations introduce diversity in the search space. To implement mutation, each and every row of the pheromone matrix is mutated with the probability P_m as follows:

$$\tau_{rs}' = \begin{cases} \tau_{rs} + \text{mut}(it, \tau_{\text{threshold}}), \text{if } a = 0 \\ \tau_{rs} - \text{mut}(it, \tau_{\text{threshold}}), \text{if } a \neq 0 \end{cases} \tag{4.7}$$

where $\tau_{threshold}$ is the average of the pheromone trail and mut($\cdot$) is a function that ensures stronger mutations with increasing counters.

4.4.4 MAX–MIN Ant System

This is another sister algorithm, in which the pheromone updation technique is varied so that the communications between the ant agents can be used in a better and enhanced way to reach solutions. It has been claimed that if the pheromone concentration increases in some part of the solution area, then it leads to a situation in which exploitation dominates exploration. So, the algorithm aims at strategic variation of pheromone updation technique so that simultaneous exploration of better solutions can be achieved through the use of stochastic methods.

To avoid stagnation of the heuristic search, in MAX–MIN ant system (MMAS), the best ant is allowed to update the pheromone trails. The word Max–Min bounds the path pheromone concentration with a maximum and minimum pheromone trail limits, which is decided by the type of problem keeping the minimum greater than zero.

Initially, the paths are set at Max and with time the pheromone trail decreases and only the best ant agents are allowed to update the pheromone concentration; and hence, the only best path survives as the iterations increase. One of the dominating forces that govern the use of MMAS is the fact that improved performance may be obtained by a stronger exploitation of the best solutions found during the search [13]. Primarily, MMAS differs from the traditional system in the following ways:

1. In MMAS, the best solutions are exploited in iteration and after each iteration, only one ant adds its pheromone.
2. In order to avoid stagnation of the search, pheromone trail ranges on each solution component are limited to an interval $[\tau_{min}, \tau_{max}]$.
3. Pheromone trails are initialized to τ_{max} so that a higher degree of exploration could be achieved.

4.4.5 Rank-Based Ant System

In rank-based ant system (RBAS), after m ants have made their tour for optimization through the search graph or the solution area, they are evaluated through their fitness values, and then ranked accordingly. Say for minimum distance optimization they are ranked $\{\lambda_1, \lambda_2, \ldots \ldots, \lambda_m\}$ where λ_m is the mth rank and λ_1 is the best rank, and then the contribution of the ant with λ_nth rank is weighted according to the rank it is evaluated [14]. This is done to reduce the *overemphasized trails caused by many ants using suboptimal paths*. In addition, in the next tour of exploration, there is a chance of balanced and considerably high exploration opportunity; and thus a better solution can be reached.

4.4.6 Ant-Q

This system incorporates reinforcement learning or precisely Q-learning [15] in which a reward is associated with every event that occurs, and optimization is achieved through the evaluation of this reward function. Let us take an example that will make the algorithm more clear and demonstrate how actually this algorithm can be applied to various applications.

Let a packet needs to be sent from one node to another node, and there is a function better known as *Reward Function* or *Evaluation Function* associated with each path. Now the packet moves and each time it moves, it gathers some rewards through how good the path is or how reliable the path is and thus establishes a new optimization technique through comparison with paths that exist between the nodes. The probability of a transition is defined by

$$p_m\{ij\} = \begin{cases} \dfrac{\left[AntQ(\text{path})\right]^{\alpha}.\left[HE(\text{path})\right]^{\beta}}{\sum_k \left[AntQ(\text{path})\right]^{\alpha}.\left[HE(\text{path})\right]^{\beta}} & \text{path} \in \Omega \\ 0 & \text{Otherwise} \end{cases} \tag{4.8}$$

where [HE(path)] is the heuristic function. α and β are parameters that control the relative importance of trail versus visibility.

4.4.7 Hyper Cube Ant System

Hyper cube ant system [16,17] is another ACO technique in which the pheromone concentration level is varied between 0 and 1 through the equation

$$\tau_{ij}(t+\delta) = (1-\rho)\tau_{ij}(t) + \sum_k \frac{[1/f(s)]}{\sum_k [1/f(s)]} \tag{4.9}$$

where $\sum_k \dfrac{[1/f(s)]}{\sum_k [1/f(s)]} = 1$ and ρ is the pheromone decay factor.

So, in vector form $\vec{\tau} \leftarrow (1-\rho).\vec{\tau} + \rho.\vec{m}$ or

$$\vec{\tau} \leftarrow \vec{\tau} + \rho.(\vec{m} - \vec{\tau}) \tag{4.10}$$

This implies that if $\vec{m}$ is the new pheromone level vector and $\vec{\tau}$ is the old level, then the updated contribution is the vector difference of $\vec{m}$ and $\vec{\tau}$, which is nothing but the third side of the triangle formed by $\vec{m}$ and $\vec{\tau}$ as the two sides. If the vectors are seen graphically, it will be seen that the points lie on a hyper plane of a unit length side cube and hence the conception and nomenclature are derived.

4.4.8 Mean-Minded Ant Colony Optimization Algorithm

A different variant of the ACO algorithm is the mean-minded ACO algorithm, which could enhance the optimized path finding criteria of the system model [5]. The main modification is done in the pheromone concentration, which will be decided by how happy is the ant at that path and not by how many ants are passing through that path for an interval of time. In the AS, all the ants would have secreted the pheromone trails and in modified versions such as BWAS, MAX–MIN, Rank-Based AS, and so on, only the suitable ant is allowed to contribute to the pheromone concentration. But in the mean-minded ACO algorithm, all the ants will contribute in different proportions following a particular function accommodating certain pertinent parameters for that ant breed. In addition, the number of ants will differ, and is decided by the mean value of the parameters of the path it has traversed. Hence, it is named ironically as *mean-minded*.

The main feature of this AS is that, only the ants having an outstanding path history or the ants developing its path parameters history will contribute in guiding the other ants and will secrete pheromone. Another important aspect of the model is that it will reject the low performing ant from the path planning. Thus, it has a clear indication of pointing out the paths in the graph, which will improve the performance and will help in quick merging of the solutions to its global best. As there are several breeds of ants, which are present in the system, there are definite pheromone types for them, which are present and each one only guides its own type without interference with the other's pheromone type. There are n kinds of ant breed being simulated, each representing one kind of optimization and there is a carefully modeled optimal cost function associated with it. So, the primary change is in pheromone concentration updation phenomenon, making it independent of time, instead of incorporating several dependent parameters in it.

Biologically, we can think it as if the ant, passing through a certain point, finds that the fitness values of the new path is less than its average (considering the parameters are set as least is better), it will secrete more pheromone, else it will secrete less. In other words, pheromone concentration will increase if the ant's optimization criteria fulfills (finding the least parametric path in the graph) and how better it is.

An example will make the model clearer. Say the ant is optimizing the cost (dependent on waiting time, distance cover time, and some rates of fuel consumption) and has traversed say m edges and is about to enter the edge$m+1$. Let CostFunction = f(WaitingTime, Speed, Distance, FuelRate{r1,r2}) and it has made X count for its cost function. So, on average it has made AvgWeight = X/m, for each edge it has covered. If the count (say X_1) is for the $(m+1)$th path and (AvgWeight< X_1), then pheromone updation will be quite lower as the ant is not happy with the path and the quality of path for optimization is below average (the quantity of parameters are above average and the function is headed for minimization).

Instead if (AvgWeight> X_1), there is a considerable amount in updation of the pheromone as the ant agent is happy with the parameter and its accumulated count. The *mean-minded* variant has provided better results in preventing stagnation of exploration criteria than the traditional algorithm and is also better in quick merging of results than the modified ones such as BWAS, MAX–MIN, Rank-Based AS, and so on.

4.4.8.1 Mathematical Formulations for Mean-Minded Ant Colony Optimization Algorithm

Let for the kth breed, the ant is optimizing with respect to some parameters $\in$ {Para$_1$, Para$_2$... ..., Para$_n$} where there are total n number of parameters. Now the cost function will be

$$\{fitness\}_k = \Psi \left\{ \left\{ \sum_m Constant * \{ParaMatrix\}_k \right\} \right\} \tag{4.11}$$

Now, the accumulated cost weight will be Sum of Para$_k$ = $\sum_m$ {Parameters}$_k$ for m paths, then for each k parameters, average will be generated, if it is divided by m. Now, we can apply the average on cost function and also the newly accumulated parameter on the cost function and we can derive the resultant pheromone accumulation estimation.

This modified version of the ACO has been solely developed for multiobjective optimization and the mathematical models also suit the states of the environment. In addition, at each iteration it is not required to judge the best and the worst or involve a ranking of the ant performance. It is admitted that at some point, it may happen that it may indulge in excessive pheromone deposition, and may hamper exploration. However, this phenomenon will help in proper adequate communication (not excessive as in Ant System and inadequate in BWAS, RBAS, MMAS, etc.) between all the ants and will help in both exploration and exploitation; hence, this clear demarcation should be paid attention and this AS will give better result than the others.

4.5 Applications

ACO algorithm has been applied to solve a variety of optimization problems across several domains. Let us briefly discuss some of the applications of the ACO as available in the literature.

A hybrid algorithm, particle ACO (PACO) has been proposed [18]. This algorithm takes advantage of ACO and particle swarm optimization (PSO) in order to solve capacitated vehicle problems. During the arch phase, (1) artificial ants intelligently construct solution routes, (2) memorize the best solution found so far, and (3) they also lay pheromone on the swarm and personal best solutions routes. To prevent being trapped within the local optima with an objective of increasing the probability of getting better solutions, PACO performs a series of pheromone disturbance and short-term memory resetting operations. These operations adjust the stagnated pheromone trails. As a result, the disturbed pheromone trails guide ants so that they can find new Pbest and Gbest solutions.

This hybrid version of the ACO algorithm proved to be very useful for solving capacitated vehicle problems.

ACO is also being extensively used in optimizing the network routing protocols; the rationale is to enhance the network life time. In [4], intercluster ACO algorithm (IC-ACO) has been proposed, which makes the use of ACO algorithm for routing data packets to and fro the network; this algorithm also makes a dedicated attempt to minimize the efforts that are otherwise wasted in transferring the redundant data that are sent in a densely deployed network. The results confirm that IC-ACO outperforms the LEACH protocol by providing higher energy efficiency, prolonged network lifetime as well as enhanced stability period.

[19] uses an ant colony meta-heuristic optimization method for solving the redundancy allocation problem (RAP). This is a well-researched NP-hard problem. This problem is usually studied in a restricted form in which each subsystem must contain identical components working in parallel; this is done to ensure that the computations are tractable. A modified ACO algorithm for solving RAP is proposed and tested on a well-known suite of problems available in the literature.

Application of Ant colony optimization on Quadratic Assignment Problem Dataset: ACO has been applied on the benchmark datasets of the well-known quadratic assignment problem (QAP) problem, and the results obtained for various parameter settings are given in the following table (Table 4.1).

TABLE 4.1

Results after Applying ACO on Standard QAP Datasets

QAP Datasets	Number of Ants	Number of Iterations	Dimensionality	Optimal	Best	Worst	Mean	Mean Error
kra32	20	500	32	88,700	89,976	94,340	91,207	2,507
had16	20	200	16	3,720	3,946	4,120	4,046	326

4.6 Summary

In this chapter, we have discussed the ACO algorithm, a very successful swarm-based algorithm in the field of optimization and scheduling-based discrete problems. The most important feature is the concrete pseudo communication between agents and the mathematics, which represent it. The algorithm lacks the operation of parameter variation. However, it still is one of the few algorithms, which are used for a wide variety of problems, including both continuous and discrete problems.

Solved Examples

1. How can ACO algorithm be used for solving TSP problem?

Solution 1

Procedure for Applying Ant Colony Optimization on Traveling Salesman Problem

In order to apply ACO on traveling salesman problem (TSP), following approach is adopted:

In each iteration, ants will be placed at random places. Each ant will complete its tour by visiting all the cities back to its starting place. The selection of next city will be based on heuristic parameter (distance) as well as the pheromone trail associated with the edges. Once the tour is completed, pheromone trails are updated for each edge. Finally, the optimal path will be the one with the highest pheromone concentration.

This process is illustrated in a step-wise manner as given in the following:

1. Initialize distance matrix
2. Initialize pheromone trail matrix
3. Initialize ACO parameters
   a. Weight of pheromone trail
   b. Weight of heuristic visibility
   c. Pheromone evaporation parameter
4. For each iteration
5. Place ants at random places
6. Repeat until tour completes
7. For each ant
8. Find the transition probability from current place to all the unvisited places using the given formula

$$P_{ij}^k(t) = \begin{cases} \dfrac{[\tau_{ij}(t)]^\alpha [\eta_{ij}]^\beta}{\displaystyle\sum_{l \in J_k(i)} [\tau_{il}(t)]^\alpha [\eta_{il}]^\beta} & \text{if } j \in J_k(i) \\ 0 & \text{otherwise} \end{cases}$$

where P_{ij} represents the probability of transition from i to j, τ_{ij} represents the pheromone trail for edge (ij), and n_{ij} is the heuristic value and α and β are parameters that control the relative importance of trail versus visibility.

9. Add the place with max transition probability to ant's tour
10. Find the best tour so far
11. Update the pheromone matrix using the following formulae:

$$\tau_{ij}(t+1) = (1-\rho)\tau_{ij}(t) + \Delta\tau_{ij}(t)$$

In this equation,

$$\Delta\tau_{ij}(t) = \sum_{k=1}^{m} \Delta\tau_{ij}^{k}(t)$$

$$\Delta\tau_{ij}^{k}(t) = \begin{cases} \dfrac{Q}{L_k} & \text{if}(i,j) \in \text{tour done by ant } k \\ 0 & \text{otherwise} \end{cases}$$

12. Find the best tour generated so far

Objective Function for Traveling Salesman Problem

Optimization function (tour length) in case of TSP with N cities is given by

$$\sum_{k=1}^{N-1} \text{distance}\left[\pi(k), \pi(k+1)\right]$$

where π (1, 2, 3.... N) represents the sequence of traversing cities.

The above-mentioned procedure has been implemented on the standard datasets of TSP and the following results are obtained (Table 4.2).

2. How should be a good ACO strategy in terms of local and global search?

TABLE 4.2
Results Obtained by Applying ACO on Benchmark TSP Datasets

TSP Datasets	Number of Ants	Number of Iterations	Dimensionality	Optimal	Best	Worst	Mean	Mean Error
berlin52	20	500	52	7542	7808	8301	7996	454
fri26	20	500	26	937	950	1023	977	40
gr17	20	500	17	2085	2103	2007	2147	62

Solution 2

In ACO algorithms, the ants have probabilities for each possible step while walking through the graph. Typically, this probability is based on two factors: (1) a local and (2) a global measure of quality. The global measure is usually associated with the pheromone deposit in an edge, and the local measure is usually related to the quality of a particular step. Therefore, if your ants are taking only greedy actions, it is possible that the probability function you are using is giving too much weight to local quality. Finding a probability function that exhibit a good compromise between the local and global search is a fundamental aspect of a successfully applied ACO strategy.

Exercises

1. Comment on the usefulness of the ACO algorithm. How does it perform better than other contemporary algorithms?

2. Discuss the inherent limitations of ACO algorithm. What steps could be taken to compensate them?

3. Why is there a need for a graph matrix in the ACO algorithm? What information does it convey?

4. How is pheromone secretion implemented in the ACO algorithm?

5. In the mean-minded ACO algorithm explain the phrase *mean-minded*.

6. Explain the rationale behind rank-based ant system.

7. What is the principle governing Q-learning?

8. Does the pheromone level affect the convergence of the ACO algorithm? Explain.

9. Explain hyper cube ant system.

10. How is best–worst ant system different from Max–Min ant system?

References

1. M. Dorigoa and C. Blum, Ant colony optimization theory: A survey, *Theoretical Computer Science*, 2005, 344(2–3), 243–278.

2. M. Dorigo, Optimization, learning and natural algorithms, Ph.D Thesis, Politecnico di Milano, Italy, 1992.

3. Y. Wan, M. Wang, Z. Ye and X. Lai, A feature selection method based on modified binary coded ant colony optimization algorithm, *Applied Soft Computing*, 2016, 49, 248–258.

4. J. Y. Kim, T. Sharma, B. Kumar, G. S. Tomar, K. Berry and W. H. Lee, Intercluster ant colony optimization algorithm for wireless sensor network in dense environment, *International Journal of Distributed Sensor Networks*, 2014, 10(4), 457402.

5. C. Sur, S. Sharma and A. Shukla, Analysis & modeling multi-breaded mean-minded ant colony optimization of agent based road vehicle routing management, *7th International Conference for Internet Technology and Secured Transactions*, London, UK, December 10–12, 2012.

6. M. Dorigo and L. M. Gambardella, Ant colony system: A cooperative learning approach to the traveling salesman problem, *IEEE Transactions on Evolutionary Computation*, 1997, 1(1), 53–66.

7. D. Gordon, *Ants at Work: How an Insect Society is Organized*, The Free Press, New York, 1999.

8. Encyclopedia Britannica. Pheromone. Britannica 2002 Deluxe Edition CD-ROM, 2002.

9. P. E. Merloti, *Optimization Algorithms Inspired by Biological Ants and Swarm Behavior*, San Diego State University, San Diego, CA, 2004.

10. M. Dorigo and T. Stutzle, Ant colony optimization: Overview and recent advances, IRIDIA—Technical Report Series, 2009.

11. K. Khurshid, S. Irteza, A. Khan and S. Shah, Application of heuristic (1-opt local search) and metaheuristic (ant colony optimization) algorithms for symbol detection in MIMO systems, *Communications and Network*, 2011, 3, 200–209. doi:10.4236/cn.2011.34023.

12. O. Cordón, I. F. Viana and F. Herrera, Analysis of the best-worst ant system and its variants on the QAP, In M. Dorigo, G. Di Caro and M. Sampels (Eds.), *Proceedings of the Third International Workshop on Ant Algorithms (ANTS'2002)*, pp. 228–234. Springer, Brussels, Belgium, 2002.

13. T. Stützle and H. H. Hoos, MAX—MIN ant system, *Future Generation Computer Systems*, 2000, 16, 889–914.
14. B. Bullnheiner, R. F. Hartl and C. Straub, A new rank based version of the ant system, *The Central European Journal of Operations Research in Economics*, 1997, 7, 25–38.
15. M. Dorigo, V. Maniezzo and A. Colorni, The ant system: Optimization by a colony of cooperating agents, *IEEE Transactions on Systems, Man, and Cybernetics—Part B*, 1996, 26(1), 29–41.
16. C. Blum and M. Dorigo, The hyper-cube framework for ant colony optimization, *IEEE Transactions on Systems, Man, and Cybernetics*, 2004, 34(2), 1161–1172.
17. C. Blum, A. Roli and M. Dorigo, HC-ACO: The hyper-cube framework for ant colony optimization, *4th Meta heuristics International Conference*, Porto, Portugal, July 16–20, 2001.
18. Y. Kao, M. H. Chen and Y. T. Huang, A hybrid algorithm based on ACO and PSO for capacitated vehicle routing problems, *Mathematical Problems in Engineering*, 2012.
19. Y. C. Liang and A. E. Smith, An ant colony optimization algorithm for the redundancy allocation problem (RAP), *IEEE Transactions on Reliability*, 2004, 53(3), 417–423.

5

Bat Algorithm

5.1 Biological Inspiration

The bat algorithm (BA) is a recent population-based meta-heuristic approach proposed by Xin-She Yang [1]. The algorithm manifests the so-called echolocation of the bats. The bats make use of SONAR echoes to detect and avoid obstacles that might be present in their path. Thus, it can be established that bats navigate by using the time delay from emission to reflection. The pulse rate is measured to be between 10 and 20 times per second, and it only remains existent for about 8–10 ms. Once the bats get hit, they transform their own pulse into useful information to bring about an exploration. The purpose of this exploration mechanism is to find out how far away the prey is? The bats usually make use of wavelengths λ that are varying in the range of 0.7–17 mm. The pulse rate can be bounded in the range from 0 to 1, where 0 means that there is no emission and 1 means that the bat's emission is at the maximum.

Following are the assumptions made regarding the behavior of the bats:

1. They use echolocation phenomenon to sense the distance, and that they are also able to distinguish between a prey and a potential barrier.
2. Bats fly randomly with a velocity v_i at position x_i with a fixed frequency f_{min}, varying wavelength λ, and loudness A_0 to search for prey, and that they can automatically adjust the wavelength of their emitted pulses as well as the rate of pulse emission.

Although the loudness can vary in many ways, we get away with the fact that the loudness varies from a large (positive) A_0 to a minimum constant value A_{min}. In essence, we consider the loudness value to be bounded (Figure 5.1) [2].

5.2 Algorithm

Steps of the Algorithm

1. The algorithm begins with an initialization (lines 1–3) step, in which we initialize the parameters of algorithm, generate, and also evaluate the initial population. In this step, we also determine the best candidate x_{best}.

Algorithm 1: Original Bat Algorithm

Input: Bat population $x_i = (x_{i1}, ..., x_{iD})$ ^T for i = 1,2, 3.... N_p,
MAX_FE

Output: The best solution x_{best} and its corresponding value $f_{min} = min(f(x))$

   1:    init_bat ();

   2:    *eval* = evaluate _the _ new _population;

   3:    f_{min} = find _the _ best_ solution (x_{best}); {initialization}

   4:    **while** termination _ condition_ not_ meet **do**

   5:     **for** i = 1 to N_p **do**

   6:        y= generate _new _ solution (x_i);

   7:        if rand (0,1) > r_i **then**

   8:         y = improve _the_ best_ solution (x_{best})

   9:        **end if** {local search step}

  10:       f_{new} = evaluate_ the_ new_ solution (y);

  11:       *eval* = *eval* + 1;

  12:       **if** $f_{new} \leq f_i$ **and** N (0,1) < A_i **then**

  13:        x_i = y; $f_i = f_{new}$;

  14:       **end if** {save the best solution conditionally}

  15:       f_{min} = find_ the _ best_ solution (x_{best});

  16:     **end for**

  17:    **end while**

FIGURE 5.1
Pseudocode for the original bat algorithm.

2. Generation of new solutions (line 6). This step witnesses the artificial bats moving within the space as per the update rules of the algorithm.

3. Local search step (step 7–9). Random walks improve on the value of the best solution.

4. Evaluation of the new solution.

5. Conditional save of the best solution (step 12–14), conditional archiving of the best solution takes place in this step.

6. Finding the best solution (step 15), the best solution of the current iteration is updated.

5.3 Related Work

Difficulties and challenges in handling problems of the real world owing to their growing abstruseness encouraged computer scientists and engineers to search for solving the problems with more efficacies. Meta-heuristic algorithms are the result of such endeavors. BA is an example of a meta-heuristic optimization algorithm that has found its utility in a series of optimization problems. Besides, different variations of BA such as chaotic BA have also been devised for solving problems such as substitution of the random number generator (RNG) supplemented by chaotic sequences for the purpose of parameter initialization. When such a variant was tested against a set of standard benchmark functions, they showed great potential for the future.

BA has also showed great potential in image reconstruction algorithms. One such application was a BAT X-ray survey [3], which was developed with an aim to retain full statistical information that is inherent in the data obtained as a result of X-ray scan and at the same time to avoid corruption of exposures, which are present in different pointing directions. In the aforementioned study, maximum likelihood (ML) method was used for image reconstruction.

An important area in image processing includes the multilevel image thresholding [4] process. This technique is used for image-segmentation process and other such procedures. However, the limitation with the classical approach is the exponential growth of the search algorithm; a phenomenon, which is certainly undesirable. However, such problems can be effectively solved by swarm intelligence methods such as BA. Such an algorithm is known to show acceptable performances when compared to other similar techniques. The existent algorithm can also be modified to impart to it elements of differential evolution and ABC algorithm with a desire to avoid getting caught in the local maxima or minima. The added functionality also ensures that a robust search is carried out in the solution phase. The following assumptions called the *ideal assumptions* are put in perspective:

1. In order to sense distances, bats usually resort to echolocation and they also have the wisdom to differentiate between a potential prey and a barrier.

2. From any given position x_i, the bats move randomly with a velocity v_i. The minimum frequency of emitted waves is a fixed frequency f_{min}. However, the wavelength can vary, and so the loudness A_0. In addition to this, it is also assumed that the bats can tune the wavelength of their emitted pulses and tweak the rate of pulse emission r as well depending on the positioning of the barrier/target.

3. Loudness is bounded between a positive large value A_0 and a minimum constant value A_{min}.

Bat optimization algorithm can also be used in power systems to solve for optimal energy management problems. This can ensure that we have in place an efficient method to supply energy to the electrical consumers, reliably and in an economical manner. In one such approach, BA has been used to solve the optimal energy management of micro-grid (MG). This unit constituted elements such as RESs [5] with the use of fuel cell (FC), wind turbines (WT), photovoltaic (PV), and micro turbine (MT) duly supplemented by storage devices to compensate for the energy mismatch. The problem is modeled with an objective of minimizing the total cost of the grid as well as that of rechargeable energy storage system (RESs). The problem also considered the interaction between MG and its utility in a 24-hour time interval. These effects were expected to aggravate the complexity with respect to principles of an optimization algorithm. The solution strategy contained a BA-based self-adaptive method to empower the BA so that it escapes the premature convergence. The simulation results confirmed that the new algorithm is not only feasible but also superior to its peers.

Bat algorithm has also been applied in the field of networking. One such area is multihop packet radio network. These networks, better known as *ad hoc* networks, are dynamic and this dynamism can be attributed to the nodes' mobility. The assignment and detachment of nodes to and from *clusters* further perturbs the topology's stability; and hence, a reconfiguration of the system becomes a necessity. However, it is equally important to stabilize the topology as much as possible. The *cluster heads* [6] form a *dominant set* in such a network and also determine its stability. The proposed algorithm takes the transmission power, node's mobility, and battery backup of the nodes, and so on as parameters. Further, the time required to identify the cluster heads is dependent on the diameter of the underlying graph. Number of nodes that can be present in a cluster has been fixed to a threshold to allow optimal operation of the medium access control (MAC) protocol.

Arranging nonidentical machines in a constrained setting of a manufacturing shop floor is an important section of plant design. In addition, material handling distance forms one of the key metrics within a manufacturing company since it leads to the efficient production at a lowered cost. Machine layout design [7] better known as facility layout problem is a subset of nondeterministic polynomial-time hard problem. The performance of BA is found to be comparable to other algorithms such as the shuffled frog leap algorithm, genetic algorithm, brain storm algorithm, and so on. This further increases the scope of BA to also include the problems in manufacturing and operations research field.

Deduplication is certainly becoming a necessity given the amount of information that is getting generated. In the existing research only genetic programming (GP) approach was shown to record deduplication. The various approaches to deduplication simply combined several different parts of substantiation extracted from the data content in order to generate a deduplication facility [8] that could detect duplication in a data depository. As record deduplication is a time-intense task even for undersized repositories, a need was felt for some sort of a faster algorithm, which does not compromise on the quality of the results obtained. As a consequence, a modified BA (MBAT) has been devised to go about checking the record duplication. The incentive behind such an approach is to generate a flexible and effective method that also has the potential to employ data mining algorithms.

The flowchart shown in Figure 5.2 describes in detail the modified algorithm. In order to supplement the results of GP, so as to find the best records and at the same time bring about deduplication (Figures 5.2 and 5.3) [8].

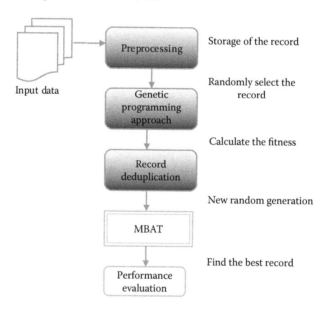

FIGURE 5.2
Flowchart describing the modified bat algorithm.

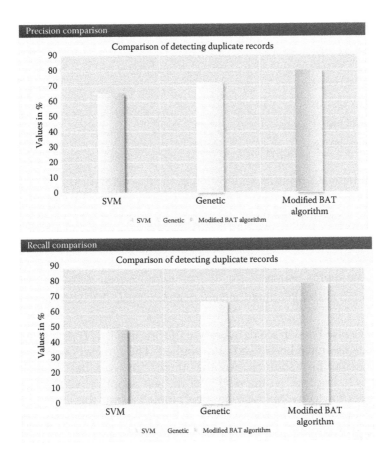

FIGURE 5.3
Comparison of MBAT algorithm with other algorithms.

Despite the success of the BA and its diverse applications, there are still a few key issues that need more research. The issues are the following:

1. Convergence rates
2. Parameter tuning
3. Large-scale problems

The standard BA works very well, but there is no well-defined mathematical analysis to link the parameters with the convergence rates. Where in principle, the convergence behavior should be controlled by the BA parameters [9], in practice, the BA converges very quickly when it is in its early stage and then the convergence rate slows down. As such, the BA is very useful to find optimal solutions to some really tough problems in a very quick time. However, the accuracy may suffer if the number of function evaluations is limited. Therefore, one of the key questions that pertain to research is "how to improve the convergence rates at the later stage?" Various methods including hybrid approaches try to improve the BA, and it would be useful to use mathematical theory to guide such research.

The different variants of BA that are prevalent in computer science are as follows:

1. Binary
2. Discrete binary bat
3. Chaotic BA
4. Cloud model bat
5. Self-adaptive bat

Due to its usefulness, BA finds its application in the following fields:

1. Multilevel image thresholding
2. Economic dispatch
3. Feature selection
4. Planning the sport sessions
5. Application in multiple unmanned combat air vehicle (UCAVs), and so on

Some efforts have also been made to hybridize the original BA. Typically, differential evolution (DE) strategies have been used and applied. These techniques have considerably improved the current best solution. Another important work has been the use of local search heuristics to improve on the results of the self adaptive bat algorithm (SABA) [2]. Domain-specific knowledge can also be incorporated using the local search. Although local search has been used in a limited way, still different DE schemes can be applied to improve the bat strategies.

Meta-heuristic algorithms [10] are considered to be higher-level methods to solve domain-specific problems. They are characterized by

1. Intensification
2. Diversification

Intensification, better known as exploitation, tends to utilize the information contained in the current best solutions. This step conducts a search around the neighborhood of the current best solutions to find the best among them.

Diversification (exploration), on the other hand, guarantees better exploration of the search space. This is usually implemented through randomization. This is an essential step as it also ensures that the system eventually comes out of any local minima.

In all such algorithms defined by stochastic components, randomness is often achieved through some probability distributions. In principle, it is more useful to replace such randomness by chaotic maps [11] since chaos have very similar properties to randomness added with better statistical and dynamical properties. Such dynamical mixing usually becomes a necessity to ensure that solutions generated by the algorithm are diverse enough to potentially reach every mode in a multimodal objective landscape. Such methods that use chaotic maps to replace random variables are known as chaotic optimization (CO) methods. Due to the ergodicity and mixing properties that chaos possesses [12], algorithms can precede iterative search steps at higher speeds than the standard stochastic search with standard probability distributions. A number of different maps can be used in this regard; for example, Chebyshev map, circle map, and intermittency maps are of significant importance to such approaches.

Swarm intelligence is also used in the realm of structural optimization [13]. In order to solve specific structural optimization, changes are required to be made to the original algorithms. Then the performance of such algorithms needs to be measured using a benchmark function. A similar work was done to optimize the truss structures to stress, stability, and displacement constraints. Typically, in a practical design optimization of truss structures the objective remains to find a minimum cost or weight design by selecting cross-sectional areas of structural members from a table of available sections, whereas at the same time ensuring that the final design meets the strength/stress and service requirements. Apparently, the robustness of the bat intelligence (BI) algorithm lies in its inherent ability to strike a perfect balance between the exploration and exploitation. The algorithm achieves this successfully by efficiently modeling the echolocation parameters during the search process itself.

The BI differs from other existing algorithms because it simulates prey hunting between the predator and the prey. The BI algorithm models the interaction between the bat and its prey repeatedly mimics such behavior mechanisms to iteratively converge toward optimal or near optimal solutions. Though many efforts have been made to solve the multiprocessor scheduling problem (single objective-based), unfortunately, limited efforts have been made toward solving such scheduling problems. A simple strategy could be to combine the objectives of multi-objective problem into a single objective using a normalized additive utility function [14]. Lexicographic ordering is one such approach in which one objective is given priority over other objectives. If the objective function value of alternatives happens to be the same as that of the most important objective, the next important objective in order determines the superior solution. Genetic algorithms (GAs) have been rigorously used to identify optimal solutions through an evolutionary approach as discussed in Chapter 3. In addition, swarm intelligence has also been developed for multi-objective optimization problems.

Following assumptions are made for the job model:

1. Tasks cannot be interrupted during its execution nor can they be killed (i.e., they are nonpreemptive).
2. Whatever be the task at hand, a single processor at hand will be able to solve it (single processor tasks).
3. Each job j_i is associated with a due date, d_i. Thus, a deadline is supposed for every task given to the system. This is done to ensure that starvation does not take place.

The multiprocessor architecture is defined by a set $P = \{p_1, p_2, \ldots\ldots, p_m\}$ of m identical processors. By identical processors, we mean that the processors have equal processing time for the same task. The processors dedicated memories and they are connected through a bus topology.

A new self-adaptive learning bat-inspired algorithm has also been devised to solve environmental/economic dispatch problem. This problem considers constraints such as valve-point effects, transmission losses, and so on, over a short-term time period [15] in a given setting. Furthermore, there is also a need to schedule sufficient resources to cater to energy and operating reserves requirements. The proposed problem is a complex nonlinear, nonsmooth and nonconvex multi-objective optimization problem. As such, its complexity further increases when the above-mentioned constraints are added. To this end, meta-heuristic bat-inspired algorithm can be used to get Pareto-optimal solutions. This algorithm also implements a self-adaptive learning procedure to increase the population diversity and in the process amends the convergence criteria.

Another important variant of original BA is *directed artificial bat algorithm* (DABA). The distinguishing features of this algorithm are as follows:

1. DABA algorithm [16] takes a generation of bats to get to the initial solution.

2. Every bat in the population has an initial position at a well-defined direction (search area scope). Initial values of these parameters are assigned in the first-generation initiation step.

3. The position value represents a potential solution, whereas the direction remains the search scope.

4. Each bat looks for a prey. During local search, the bat uses directed search scope to search neighborhood solutions.

5. The artificial bats in DABA wisely mimic the behavior of bats in the real world. The wave frequency is increased only in case the bat happens to find a prey; otherwise, it is decreased. Taking into account the fact that the product of wavelength and frequency remains a constant, the following curve was obtained when traveling salesman problem (TSP) was solved using DABA.

The results confirm that DABA gives better solutions for the same number of iterations. In addition, there is a 4% to 9% fitness value enhancements compared to artificial bee colony (ABC) algorithm.

Another application of swarm intelligence algorithms is in the modeling of a conventional power system stabilizer (CPSS). The objective is to optimize its gain and pole-zero parameters. For this, the standard CPSS parameters are considered. The objective function is based on the eigenvalue shifting in order to guarantee the stability of a wide range of operating conditions. In this application, BA-CPSS algorithm outperformed PSO-CPSS algorithm.

Usually, CPSS structures [17] are modeled as lead-controller (single-stage CPSS), lead–lag controller (double-stage CPSS) with three stages of operation. These stages may include light, nominal, and heavy-loading conditions. These stages are provided to tune the parameters of the CPSS model. The computational time and burden get increased with an increase in the size/load of the controller. Taking into account the fact that the operating condition of the power system is dynamic, a robust controller should be designed and it should be tested over a wide range of conditions.

Attribute reduction (AR) [18] is a problem of choosing an optimal subset of attributes from its superset to build better prediction models. This forms a subset of feature engineering and is extensively being researched not only in the domain of machine learning but also in data analytics and signal processing. To efficiently bring about the attribute reduction, a variant of bat by the name, BA for attribute reduction (BAAR) has been proposed. The driving forces behind attribute reduction are the following:

1. To increase the predictive accuracy of the algorithms

2. To reduce the dimensionality of the feature space

3. To pick up the comprehensibility among the feature vectors

4. Visualization of the induced concepts

5. To simplify the models in the given complex real-world data

Although BA is known for its simple design and easy implementation, it suffers from premature convergence; a characteristic found in majority of the stochastic optimization algorithm. There are some inherent deficiencies in BA:

1. Each component present in the bat's flight velocity uses the same frequency increment; as such, the flexibility of the motion is constrained. As a consequence, most of the bats will be unable to avoid a hostile attack and at the same time it also runs the risk of getting immersed into the blind alley. Algorithmically, it is equivalent to getting into the local extremum.
2. The random update in the position of the bat is not considered in the original algorithm. This causes the bat to move to the position of its prey without considering the fact that the position of the prey might have changed.

An improved BA called bat algorithm with recollection (RBA) [19] is implemented to do away with the limitations of BA. Experiments conducted on the sphere and Rastrigin function suggest that the algorithm with *recollection* ability shows faster convergence rates as compared to the original algorithm (Figure 5.4) [19].

A novel robust hybrid meta-heuristic optimization approach has been devised in order to solve optimization problems that are global in nature. The enhancements include the incorporation of pitch adjustment step in a harmony search algorithm. This search algorithm models mutation operator when the position of the bat is getting updated. This approach, thus, helps to speed up the convergence rate and makes the algorithm more

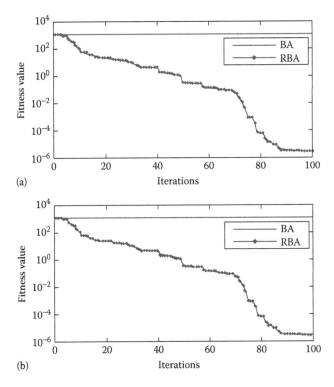

FIGURE 5.4
Graph showing the performance of the algorithm for (a) sphere and (b) Rastrigin function.

Begin

 Step 1: Set the parameter and initialize the HM.

 Step 2: Evaluate the fitness for each individual in HM.

 Step 3: While the halting criteria is not satisfied do

   for d = 1: D **do**

    if rand < HCMR **then**   //memory consideration

     x_{new} (d) = x_a (d) where a $\in$ (1,2,3......., HMS)

     if rand < PAR **then**   //pitch adjustment

      x_{new} (d) = x_{old} (d) + bw × (2 × rand -1)

     end if

    else   //random selection

     x_{new} (d) = $x_{min, d}$ + rand × ($x_{max, d}$ -x_{min-d})

    end if

    endfor d

    update the HM as x_w = x_{new}, if f (x_{new})< f (x_w) (minimization objective)

    update the best harmony vector found so far

 Step 4: end while

  Step 5: Post-processing results and visualization.

 End

FIGURE 5.5
Pseudocode for the hybrid bat algorithm. HM: Harmony Memory; HCMR: Harmony Memory Consideration Rate and PAR: Pitch Adjustment Rate. (Reproduced from courtesy of The Electromagnetics Academy. With permission.)

amenable to a large subset of real-world applications [20]. The pseudo code of the hybrid bat algorithm is shown in Figure 5.5 [20].

The above-mentioned approach has also been tested on 14 different benchmark functions. The simulation results confirmed that the performance of (HS/BA) is superior to, or at least comparable to, that of optimization methods, such as ACO, BA, PSO, and so on, (Figure 5.6) [20].

Another variant of BA uses a differential operator in order to accelerate the convergence process. The approach is very much similar to the one used by *DE/best/2* in differential variant. Levy flights trajectory ensures not only that the randomness (diversity of the population) is maintained but also that there are minimized chances of premature convergence. This ensures that the algorithm jumps out of local minima. The algorithm was tested against 14 standard benchmark functions and the results showed its dominance in providing better convergence speeds. The distinguishing characteristics of this algorithm include the following:

1. *Stability*: Distribution of sum of independent random variables is same as that of each variable.

2. Power law is asymptotic in nature (*heavy tails*).

3. Generalized version of central limit theorem holds good.

In differential operator and Levy flights trajectory (DLBA) [21], the frequency goes for a large swing and can self-adaptively change. In addition, a differential operator is also introduced. This mimics the mutation operator present in the DE (Figure 5.7) [21].

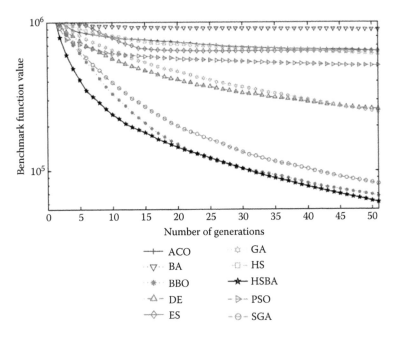

FIGURE 5.6
Performance comparison of different methods using F02 Fletcher Powell function.

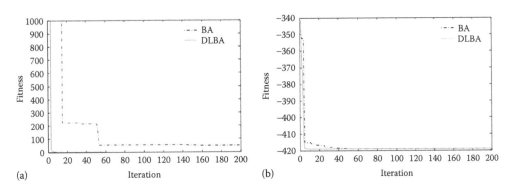

FIGURE 5.7
(a) Simulation results for Schwegel's problem and (b) Schwegel's function.

Solved Examples

1. In what ways are conventional BA better than other swarm intelligence algorithms?

Solution 1

The better performance of BA is attributed to the following:

1. BA uses echolocation and frequency tuning to bring about optimization. Though these phenomena are not directly mimicked, still they go a long way in providing extra functionalities to the algorithm.

2. *Automatic zooming*: By virtue of this property, BA has a distinct capability to automatically zoom into a region simulating exploitation in the process. This, in addition, is supplemented by an automatic switch from exploration-to-exploitation. As a consequence, BA converges quickly.

3. BA, in contrast to conventional swarm intelligence algorithms, uses parameter control. This enables the algorithm to vary the parameter (A and r) values as the iterations proceed.

2. A bat wishes to prey on its target whose size is of the order of 15 mm. Calculate the frequency of waves that it should emit for the same. Assume the speed of sound v to be 340 m/s.

Solution 2

Given

Size of the prey = 15 mm

That is, wavelength of the emitted waves should be 15 mm

Velocity of sound in the air = 340 m/s

Hence, using $v = f.\lambda$, we have,

$F = v/\lambda$

Frequency of waves that need to be emitted = 340 m/s/(15 × 10⁻³ m) = 22.6 KHz

3. Find the wavelength of ultrasonic sound waves emitted by a bat given the frequency of emitted waves to be 35 KHz. Assume velocity of sound to be 340 m/s. It is given that the bats are emitting 200 waves per second.

Solution 3

Given

Velocity of sound in air (v) = 340 m/s

Frequency of the waves emitted (f) = 35 KHz

Therefore, the wavelength of the emitted waves (λ) = v/f

$= 340 \text{ m/s}/(35 × 10^3\,\text{s}^{-1})\text{c}$

$= 9.7 \text{ mm}$

Exercises

1. Discuss the physical significance of pulse rate and loudness.
2. How are pulse rate and loudness manifested in BA?
3. Discuss the consequence of simulated annealing in BA.
4. What are the major limitations in using BA for problems concerned with power systems and control engineering?

5. Discuss the role of acceptance probability in the convergence and accuracy of the solutions obtained.

6. How is self-adaptive variant of BA different from the original algorithm? Discuss.

7. Discuss the importance and consequence of attribute reduction.

8. What is the physical significance of bat echolation process?

9. Discuss the notion of *chaos* in Chaotic Bat algorithm.

References

1. X. S. Yang, A new metaheuristic bat-inspired algorithm. *Nature Inspired Cooperative Strategies for Optimization (NICSO 2010)*, 2010, 65–74.

2. A. H. Gandomi, X. S. Yang, A. H. Alavi and S. Talatahari, Bat algorithm for constrained optimization tasks, *Neural Computing and Applications*, 2013, 22, 1239–1255.

3. M. Ajello, J. Greiner, G. Kanbach, A. Rau, A. W. Strong and J. A. Kennea, Bat X-Ray survey. I. methodology and X-Ray identification, *The Astrophysical Journal*, 2008, 102–115.

4. A. Alihodzic and M. Tuba, Improved bat algorithm applied to multilevel image thresholding, *The Scientific World Journal*, 2014.

5. M. Chatterjee, S. K. Das and D. Turgut, WCA: A weighted clustering algorithm for mobile Ad hoc networks, *Cluster Computing*, 2002, 5, 193–204.

6. K. Dapa, P. Loreungthup, S. Vitayasak and P. Pongcharoen, Bat algorithm, genetic algorithm and shuffled frog leaping algorithm for designing machine layout.

7. A. F. Banu and C. Chandrasekar, An optimized approach of modified BAT algorithm to record deduplication, *International Journal of Computer Applications*, 2013, 62, 10–15.

8. I. Fister, I. Fister, X. S. Yang, S. Fong and Y. Zhuang, Bat algorithm: Recent advances, *IEEE International Symposium on Computational Intelligence and Informatics*, 2014, 15, 163–167.

9. I. Fister, S. Fong, J. Brest and I. Fister, A novel hybrid self-adaptive bat algorithm, 2014.

10. A. H. Gandomi and X. S. Yang, Chaotic bat algorithm, *Journal of Computational Science*, 2014, 5, 224–232.

11. O. Hasançebi, T. Teke and O. Pekcan, A bat-inspired algorithm for structural optimization, *Computers and Structures*, 2013, 128, 77–90.

12. H. Afrabandpey, M. Ghaffari, A. Mirzaei and M. Safayani, A novel bat algorithm based on chaos for optimization tasks, 2014.

13. T. Niknam, R. A. Abarghooee, M. Zare and B. B. Firouzi, Reserve constrained dynamic environmental/economic dispatch: A new multi-objective delf-adaptive learning bat algorithm, *IEEE Systems Journal*, 2013, 7(4), 763–766.

14. A. Rekaby, Directed artificial bat algorithm (DABA), *International Conference on Advances in Computing, Communications and Informatics*, Mysor, India, 2013, pp. 1241–1246.

15. D. K. Sambariya and R. Prasad, Robust tuning of power system stabilizer for small signal stability enhancement using meta-heuristic bat algorithm, *Electrical Power and Energy Systems*, 2014, 61, 229–238.

16. A. M. Taha and A. Y. C. Tang, Bat algorithm for rough set attribute reduction, *Journal of Theoretical and Applied Information Technology*, 2013, 51(1), 1–8.

17. G. Wang and L. Guo, A novel hybrid bat algorithm with harmony search for global numerical optimization, *Journal of Applied Mathematics*, 2013, 2013, 21.

18. J. Xie, Y. Zhou and H. Chen, A novel bat algorithm based on differential operator and lévy flights trajectory, *Computational Intelligence and Neuroscience*, 2013.

19. A. Baziar, A. Kavoosi-Fard and J. Zare, A novel self adaptive modification approach based on bat algorithm for optimal management of renewable MG, *Journal of Intelligent Learning Systems and Applications*, 2013, 5, 11–18.

20. B. Malakooti, H. Kim and S. Sheikh, Bat intelligence search with application to multi-objective multiprocessor scheduling optimization, *International Journal of Advanced Manufacturing Technology*, 2012, 60, 1071–1086.

21. W. Wang, Y. Wang and X. Wang, Bat algorithm with recollection, *Progress in Electromagnetics Research*, 2007, 207–215.

6

Cuckoo Search Algorithm

6.1 Introduction

This chapter deals with the relatively new algorithm, that is, cuckoo search (CS) algorithm proposed by Deb and Yang in 2009 [1], and it has been found to be competent and capable in dealing with problems that need to be globally optimized. The optimization is increasing hastily and is far more important as time, resources, and money are limited. The basic objective of optimization is to accomplish a stable system. Today, in approximately all areas of industry and science we are working to achieve a stable solution or we can say we are working to optimize something, that is, whether to minimize the resource requirement or to maximize the profit. In the same context, many algorithms inspired by nature are emerging and are used extensively for solving these wide ranges of problems dealing with optimization.

CS is a novel nature-inspired meta-heuristic algorithm, aiming to obtain convergence for continuous optimization problems more rapidly similar to other nature-inspired algorithms. The algorithm is rooted on a fascinating reproduction technique of cuckoo birds. These birds never make nests of their own. Rather, they rely on other host bird's nest for laying their eggs. Most surprisingly, cuckoo eggs mimic physical characteristic of host bird's eggs in context of spot and color. If their strategy of hiding their eggs is unsuccessful, that is, the host bird discovers cuckoo's eggs, there are two possibilities. First, it can dump the eggs away or second, find some other place after leaving the nests.

The nests are initialized randomly for laying eggs. Each nest contains an egg n. The number of eggs in each nest is one in a simpler one-dimensional problem but increases according to the change in dimensions. Eggs having better fitness are passed to the next generation; on the other end, eggs having low fitness are replaced with the probability p_a.

One of the prominent features that distinguish CS from other algorithms is the concept of levy flight. Analogous to the random walk performed in bat algorithm (BA) for generation of new position, CS uses a complex levy flight [1] random walk strategy that places CS ahead of other optimization techniques. A research by Frye and Reynolds [2] shows that fruit flies or *Drosophila melanogaster*, unearth the area of exploration using a series of straight flight paths followed by a rapid turn of 90°, leading to a free search pattern of a levy-flight-style. A walk is performed in which the steps are described in terms of step length with directions of steps being random and isotropic determined by certain probability distributions. Figure 6.1 depicts the levy flight.

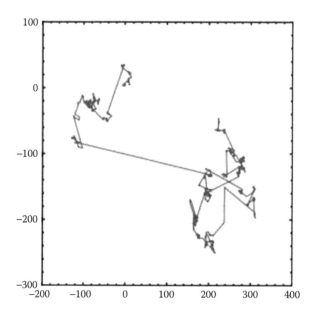

FIGURE 6.1
Example of levy flight.

6.2 Traditional Cuckoo Search Optimization Algorithm

CS algorithm can ideally be described by considering these three rules:

- Each cuckoo lays eggs in randomly chosen nests.
- High-quality eggs contained in some nests are passed over to the next generation.
- The host nests is limited to some fixed number, with a possibility $p_a \in [0,1]$ that eggs are discovered by host and those can either be discarded or may result in generation of a new nest.

Considering the three rules mentioned above, CS can be described algorithmically as follows:

1. Generate a population of n eggs randomly.
2. Get a cuckoo by levy flight [1] random walk strategy.

$$x_i(t+1) = x_i(t) + \propto \ \oplus \text{Levy}(\lambda) \qquad (6.1)$$

3. Evaluate the fitness of the randomly chosen cuckoo.
4. Compare the fitness of a new solution with an old solution. Keep the best eggs as new solution.
5. Fractional worst eggs are abandoned with a probability P_a and generate new ones through the levy flight.
6. Keep the best solutions.

7. Repeat steps (2)–(6) till the iteration exhausts.

8. The best nest with best objective function is determined.

Figure 6.2 represents the flowchart of the above-described algorithm. Unlike other algorithms, CS requires the adjustment of only two parameters. One is the levy flight coefficient α and the second is the probability p_a in reference to which the eggs are discarded. In maximization problems, the fitness function depends directly on the cost of objective function. For implementing CS, let us assume that each nest's egg represent a solution. The eggs with lesser fitness values are replaced by performing a levy flight.

$$x_i(t+1) = x_i(t) + \alpha \oplus \text{Levy}(\lambda) \qquad (6.2)$$

Here α as mentioned earlier is the levy flight coefficient also known as the step-size scaling parameter. In most cases, we can use α = 1. Random walk performed in levy flight is a Markov chain that depends only on the current state regardless of the last known

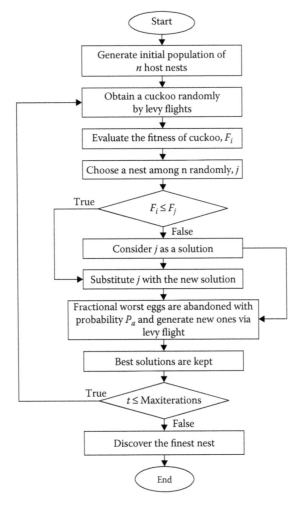

FIGURE 6.2
Flowchart of cuckoo search algorithm.

states. The XOR operator used represents the entry-wise multiplication and is more efficient because of the longer step lengths. The levy distribution is used for calculating the step length and the same is represented as in Equation 6.3:

$$\text{Levy}\left(u\right) = t^{-\lambda} \quad \text{where } 1 < \lambda \leq 3 \quad (6.3)$$

The distribution has an infinite mean with an infinite variance. With the random walk as specified, the new solutions are generated, and thus works around the best solution obtained, which in turn speeds up the local search. The above-mentioned algorithm can also be transformed into a pseudocode. Hence, the pseudocode for CS algorithm is

---

Start
Objective function $f(x)$, $x = (x_1, x_2 ... x_u)^T$
Generate initial population of n host nests x_i ($i = 1, 2, 3 ... .n$)
While ($t <$ maxGenerations) or (stop criterion)
    Get a cuckoo randomly by Levy flights
    Evaluate its quality (fitness) F_i
    Randomly choose nest among n available nests (say j)
    If ($F_i > F_j$)
        Replace j by new solution;
    End
    A fraction (p_a) of worse nests are abandoned and
    New nests are build;
    Keep the best solutions or nests with quality solutions;
    Rank the solution and find the current best
End while
Post process and visualize results
End

The random walk based on the levy flight and fewer numbers of parameters as compared to other bio-inspired algorithms is the outstanding feature of CS. Walton et al. [3] in 2011 further improved the convergence speed of the algorithm. He proposed a modification of the algorithm that modification involves the exchange of information between the top eggs.

Another change in the existing approach is the modification of step size. He tested the robustness of his proposed method on both unimodal and multimodal functions as well as on varying dimensions. His results indicated that modified cuckoo search (MCS) is well suited to problems having large number of attributes.

6.3 Variants of Cuckoo Search Algorithm

Several variations of the CS optimization algorithm have been proposed so far, using which optimizations have been performed considering both single as well as multiple objectives. In this section, we will discuss various improved/modified variants of the traditional CS algorithm and their features will be illustrated. We will also discuss the performance of the algorithm for various applications.

6.3.1 Modified Cuckoo Search

The MCS algorithm proposed by S. Watson et al. [3]. We had implemented it on gait database. The basic steps involved in the MCS algorithm are as follows:

1. *Generate initial population*: Randomly generate a population of n eggs. In the binary version of this algorithm, search space can be analyzed as n-dimensional Boolean lattice, in which the best fitness lies across the corners of a hypercube. A binary vector similar to bat algorithm (BA) is employed, where "1" represents presence of feature and "0" represents absence of a particular feature.
2. Initialize the best fit and global fit to a minimum value.
3. *Evaluate population*: At every iteration, the population is evaluated for best solution. After that, the population is scaled, so as to get the maximum fitness value by selecting the minimum number of features.
4. Sort eggs by the order of fitness.
5. Replace the worst eggs with new solutions. Here, the step size $A \leftarrow A/\sqrt{G}$, where G: current iteration number.
6. Crossover of top eggs is performed, so as to converge quickly to the best solution. Step size A varies as $A \leftarrow A/G^2$.
7. Repeat steps (4)–(6) till condition exhausts.

Figure 6.3 represents the flowchart of an implemented MCS algorithm. In every iteration, we evaluate the set of eggs in each nest. Similar to BA, a string "10011" represents a five-dimensional problem in which the 1st, 4th, and 5th eggs are considered to determine the fitness. At each iteration, the fitness of population is evaluated, so as to update the global best. The nests are then ranked according to fitness and a fraction of nests (here $p_a = 0.25$) are regenerated in the hope of getting a more feasible solution. After that, for each top egg, a random egg is drawn from the elite of top eggs and information is exchanged between the two. In other words, a new egg is generated from these two elite eggs by performing a random walk and is included in the population if new egg's fitness is better than the two.

The unique feature of CS is to discard the worst eggs and cross-breeding of elite eggs to produce an efficient solution that separates it from other meta-heuristic approaches. In addition, another benefit of CS is the implementation of levy flight rather than standard random walk techniques. As levy distribution has infinite variance and mean, MCS can see the sights or the space more efficiently when matched with other approaches. Yang et al. [4] proved mathematically that CS converges globally more efficiently.

6.3.2 Improved Cuckoo Search Algorithm with Adaptive Method

Though CS algorithm has the ability to always get to the optimal solution, still the improvement of CS involves improving the convergence rate of the algorithm. As discussed, in a traditional CS algorithm, levy flying step length is a constant, that is, α is a constant, which, in practice, reduces the adaptive method in an engineering problem. Consequently, Zhang and Chen in 2014 [5] proposed an improvement on the traditional

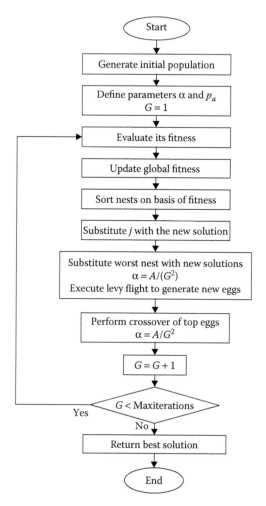

FIGURE 6.3
Modified cuckoo search flowchart.

CS algorithm, which involves a change in the value of α, for every iteration. Thus, the improved CS algorithm uses an adaptive method for calculating the step-size scaling parameter (α) and is defined as

$$\alpha = \frac{e^{\frac{\ln \alpha_{\min} - \ln \alpha_{\max}}{N_{\max}} \times N}}{\sqrt[4]{N}} \tag{6.4}$$

where $\alpha_{\max}$ and $\alpha_{\min}$ signify the maximum and the minimum of step length, respectively, and N is the iteration times of this method with initial value 1, $N_{\max}$ is the maximum of N.

As the generation becomes larger, the levy flying step length becomes less with this change in the value of α. The better performance has been achieved on various test functions with improved CS; improved CS (ICS) algorithm converges with higher speed and precision as compared to the traditional CS algorithm. ICS algorithm incorporates the advantages of CS algorithm, such as faster speed, easy to implement, and has fewer number of parameters.

6.3.3 Multiobjective Cuckoo Search Algorithm for Design Optimization

In today's era, most of the problems in engineering for design optimization are typically multiobjective. We often need to optimize a problem considering multiple objectives. The algorithms required for solving single objective problems can be considerably different from the methods for multiobjective optimization problems [6,7]. For multiobjective optimization the cost of computing, the number of parameters, and the number of function assessments may be very large. Many meta-heuristic algorithms inspired by nature have begun to show their importance in dealing with multiobjective optimization. In fact, these algorithms are exceptionally influential in dealing with this sort of problems. In this section, optimization considering multiple objective problems using the CS algorithm has been discussed.

Multiobjective problems are often difficult and time consuming as compared to single objective problems. In this type of problem, we have to generate multiple optimal solutions and, which in turn, increases the computational cost depending on the problem complexity.

We have different algorithms for single objective and multiple objectives problems when considering the implementation point of view. Algorithm that works fit for one case, that is, single objective optimization need not fit for the other case, that is, for multiobjective problems, neglecting certain unusual state of affairs such as converting multiobjectives into a single objective using some method. Some changes or modifications are to be incorporated, so as to make the algorithm also fit for multiobjective problems. In addition, algorithm should be capable of dealing with real-world problems because there is some degree of uncertainty or noise that is associated with real-world optimization problems.

Regardless of these challenges, we have a successful application of algorithms dealing with multiobjective problems. In the same respect, meta-heuristic algorithms such as CS inspired by nature have shown their worthiness in solving these kinds of problems.

We have discussed the three rules that are considered for explaining the traditional CS algorithm in section 6.2.

For dealing with multiple objectives, that is, multiobjective optimization problems, Yang and Deb [8] in 2013, proposed a modification in the first and third rule:

- K eggs are laid by each cuckoo at a time, and dumps them in an arbitrarily selected nest. Egg k corresponds to the solution to the kth objective.
- Worst nests are abandoned with a probability P_a and according to the similarities or differences of the eggs a new nest with K eggs will be built. To have a diverse solution some random mixing can also be used.

Where, K is the number of different objectives to be considered for optimization. To generate a new solution randomly by a random walk (levy flight), we can transform the first rule into a randomization process. At the same time, crossover is performed by a localized random permutation over solutions. At most, there can be K solutions, which are generated for each nest in the same way as Equation 6.1. Furthermore, based on the survival of the fittest theory, the fittest solutions are selected for the next iteration using the second rule, and this selection of the fittest helps to make sure that the algorithm converges efficiently. Further, the third rule can be well thought out as same to the mutation, which in result discards the worst solutions with a probability and also new solutions are generated, according to the resemblance of solutions with other solutions. Thus, the improved efficiency of the algorithm is ensured with these unique features [8].

The general steps of the multiobjective cuckoo search (MOCS) can be briefly transformed as the pseudocode:

Start
Objective function $f_1(x),.....,f_k(x)$ $x =(x_1,x_2...x_d)^T$
Generate initial population of n host nests x_i and each with k eggs
While (t< maxGenerations) or (stop criterion)
    Get a cuckoo randomly by Levy flights
    Evaluate its quality (fitness) and check if it is a Pareto optimal
    Randomly choose nest n (say j)
    Evaluate K solutions for nest j
    If new solution of nest j dominate those of nest i,
        Replace nest i by the new solution set of nest j;
    End
    A fraction (p_a) of worse nests are abandoned and
    Keep the best solutions or nests with quality solutions;
    Sort and find the current Pareto optimal solutions
End while
Post process with visualization of results
End

6.3.3.1 Pareto Front

The Pareto front (PF) of a multiobjective can be defined as the set of nondominated solutions so that

$$PF = \{s \in S | \exists s' \in S : s' < s\}$$

Or in terms of Pareto optimal set in the search space

$$PF^* = \{s \in f | \exists s' \in f : f(x') < f(x)\}$$

Where $f = (f_1,.....,f_k)^T$

In addition, a dominating relationship is defined as where no component of $u = (u_1,.....,u_n)^T$ is larger than the corresponding component of $v = (v_1,....., v_n)^T$, and at least one component is smaller. We can define the dominating relationship $\leq$ by

$$u \leq v \rightarrow u < v \text{ or } u = v$$

The above-mentioned algorithm considering multiobjectives for optimization using CS has been applied to solve the design optimization benchmark problems after proving its importance against well-known test functions.

6.3.4 Gradient-Based Cuckoo Search for Global Optimization

A.B. Petriciolet and Fateen [9] in 2014 stated that the mathematical performance of stochastic optimization methods can be enhanced by using gradient in conventional CS algorithm. With the use of the gradient we can improve the performance and accuracy finding of the global optimal solution. In this section, we will study an additional modification to the CS algorithm using gradient. The use of gradient in conventional CS algorithm improved the consistency and usefulness of the algorithm. In addition, optimization problems solved by using gradient-based cuckoo search (GBCS) have proved it to be a strong contender to other algorithms.

The authors incorporate knowledge about the gradient of the objective function by introducing the simple adaptation to the conventional CS algorithm. The performance and the nature of the algorithm should not be affected by any modification to the algorithm. A local arbitrary or random walk is modified in which a fraction $(1-p_a)$ of the nests is replaced.

$$x_i^{t+1} = x_i^t + \alpha \left(x_j^t - x_k^t \right) \tag{6.5}$$

Where x_j^t and x_k^t are selected randomly by random permutation as two different solutions and α is a random number drawn from a uniform distribution.

In the traditional CS algorithm, the magnitude and the direction of the step are both random, when the worst nests are substituted by new nests through a random step. Hence, in the modified algorithm, the direction is calculated based on the sign of the gradient of the function. The step direction is made positive if the gradient is negative, the step direction is made negative if the gradient is positive. Thus, new nests are found randomly from the worse nests but in the direction of the minimum as seen from the point of view of the old nests [9]. On the other end, this modification in the algorithm keeps the step magnitude as random step as same as the traditional CS algorithm. Thus Equation 6.5 is replaced as

$$\text{step}_i = \alpha \left(x_j^t - x_k^t \right) \tag{6.6}$$

$$x_i^{t+1} = x_i^t + \alpha \left(x_j^t - x_k^t \right) \oplus \text{sign} \left(-\frac{\text{step}_i}{df_i} \right)$$

Or

$$x_i^{t+1} = x_i^t + \text{Step}_i \oplus \text{sign} \left(-\frac{\text{step}_i}{df_i} \right) \tag{6.7}$$

Where, sign function obtains the sign of its argument and df_i is the gradient of the objective function at each variable, that is, $\partial f / \partial x_i$.

The overall structure of the CS algorithm is reserved with the use of a gradient in the traditional CS. Furthermore, available information is utilized for changing the gradient of the objective function. To implement this change, no additional parameter is required.

6.4 Applications

CS has proved its worth in various fields of problems on computational intelligence, and optimization with an improved efficiency. For instance, in the domain of electrical engineering, manufacturing engineering, computer engineering, design applications, and many more, CS has improved its efficiency and achieved better performance over other algorithms for various optimization problems.

In addition, a modified CS by Walton et al. [3] has demonstrated to be very efficient for solving nonlinear problems such as mesh generation. Vazquez [10] used CS to train spiking neural network models, whereas Kaveh and Bakhshpoori [11] used CS to successfully design steel structures, Kumar and Chakarverty [12] gained optimal design for reliable embedded system using the CS algorithm, and Chifu et al. [13] used CS for clustering food offers. Furthermore, Yildiz [14] has used CS to enhance the results in selecting the most favorable machine parameters in a milling operation.

CS algorithm has also been applied in the fields of wireless sensor networks (WSN) [15], data fusion in WSN [16], to solve well-known knapsack problem [17], and to solve scheduling problem for nurses [18]. CS algorithm has also been applied in gaming innovations [19]. In addition, CS has been used in field of medical science, for the recognition of Parkinson's disease.

6.4.1 Recognition of Parkinson Disease

Parkinson's disease is one of the most frequent degenerative disorders of the central nervous system. It affects the nerve cells in the brain that produces dopamine. The major symptoms include muscle rigidity, akinesia, tremors, and significant changes in speech and gait. Parkinson's disease is a chronic neurological disorder that directly affects the human gait. Parkinson's disease causes slow body movement. Early recognition of the disease is very important and for this gait serves as a major outcome measure. Using gait techniques, we had performed a comparative analysis of nature-inspired CS algorithm for an effective and quick recognition of Parkinson's disease. Recognition of disease involves selecting optimal or favorable features, which can classify the affected patients from others. The successful implementation of modified CS algorithm (Section 6.3.1) for optimal feature selection is done. This study not only provides a theoretical aspect of gait recognition but also provides a practical insight thereby radically improving the chances of early detection in real life.

6.4.2 Practical Design of Steel Structures

Recently, meta-heuristic algorithms have been implemented to surmount the rigid nature of optimum design of structures. CS algorithm emerged as a type of population-based algorithm that has found a prominent position in optimizing the design of structures of steel. The objective for optimization is reducing the self-weight of real-size structures to a minimum. The same is achieved by an appropriate selection of sections from the American Institute of Steel Construction (AISC) wide-flange (W) shapes list. The objective of a CS algorithm is to optimize self-weight of three inclusive steel structures with the same structural system on two perpendicular directions. Strength constraints of the AISC load and resistance factor design specification, geometric limitations, and displacement constraints are imposed on frames.

Based on this study, the authors [11] concluded that the CS algorithm gives acceptable convergence speed and performance from the early iterations only.

6.4.3 Manufacturing Optimization Problems

Complex nonlinear optimization problems can be effectively tackled with the meta-heuristic algorithms [20]. Among various applications CS algorithm also finds its application in solving manufacturing optimization problem. CS algorithm has been applied for the optimization of machining parameters in the literature.

In machining applications, basic objectives to be considered are maximizing the rate of production, minimizing the cost of operation, and also to improve the quality of machining. The basic objective here is to produce a product of high quality and with minimum cost. The same can be attained through optimization techniques.

Yildiz [14] has implemented CS algorithm in a milling operation by selecting optimal machine parameters. The cutting parameters in milling operations can be optimized using a CS algorithm. The application of CS for optimization of parameters in a milling operation has given enhanced results.

Although various other optimization techniques have been incorporated for the same, but the results obtained signifies the superiority of CS over other algorithms such as genetic algorithm (GA) [21], ant colony algorithm [22], and many other.

Other possible applications of the CS can be on other metal-cutting problems such as turning, grinding, drilling, operations for design optimization in manufacturing industry [14], and so on. Furthermore, CS can also solve designing heat exchangers in engineering applications.

6.4.4 Business Optimization

Global search can be effectively carried out by using various meta-heuristic algorithms. In the same discipline, CS has evolved as a popular and powerful algorithm for global optimization. Many studies have concentrated on neural networks and support vector machine for business optimization.

Business optimization often concerns with massive but often incomplete datasets, evolving dynamically over time. Some tasks cannot start before other required tasks are completed, such complex scheduling is often NP-hard, and no universally efficient tool exists. Recent trends indicate that meta-heuristics can be very promising, in combination with other tools such as neural networks and support vector machines [23].

In the same context, the use of CS to carry out optimization tasks has been achieved and performance comparison of CS with a support vector machine has also been done. By testing benchmarks problems of business optimization such as project scheduling and bankruptcy predictions, authors [23] conclude that CS can perform superior than a support vector machine.

6.4.5 Optimized Design for Reliable Embedded System

Most of the problems in the real world are multiobjective optimization problems. As we have already discussed the multiobjective CS algorithm for design optimization, we are now discussing the application of the same for designing reliable embedded systems. Multiple objectives that are to be considered for optimizations are reliability, availability, cost, performance, and so on. Dealing with the multiple objectives at the same time is definitely not an easy task as these design objectives may conflict with one another.

Say, improving the availability requires either to employ redundant components or to use highly dependable components, but both may increase the cost objective at the same time. So, in order to deal with these conflicting multiple objectives we can formulate a rather simple approach, that is, we consider one of these parameters as a design objective and model other parameters as design constraints. CS optimization algorithm can be used to solve the multiobjective reliability optimization problem and furthermore, for the optimization of scheduling and allocation for reliable embedded system.

After successful demonstration, authors [12] came to the conclusion that CS is less sensitive to variations in tuning parameters. In addition, algorithm gives improved solutions and helps in building progressively on good solutions found during the process of design exploration. Hence, CS has proved its advantages for an optimized design in embedded systems.

6.4.6 Face Recognition

Face recognition includes processing the query image by extracting the features, that is, feature selection and then comparing the image with the most ballpark (approximate) images stored in the database. First, selection of features of images stored in the database is done and then an unknown face is recognized by matching these features with the features of a known image. Now, optimization techniques can be applied in feature selection process. Optimization involves removing the noisy, irrelevant, and redundant data during the feature selection process and consequently leading to an improved accuracy in face recognition. Hence, CS algorithm can be used for selecting the optimal features, that is, in the process of face recognition.

Face recognition involves feature extraction and then feature selection. For extracting features from image discrete cosine transformation can be used. Now, as features are extracted next job is to select the relevant features from the many extracted features. Optimization using CS aims in selecting the most relevant features from many extracted features, and thus the optimized feature subset obtained contains the most approximate or representative subset and is used to recognize the face from the database having face images. The idea of applying the cuckoo algorithm (pseudocode) for the feature selection is presented as follows:

1. *Feature extraction*:
   Apply DCT to an image and obtain the DCT array
2. Select the most significant features of size $n \times n$ in the upper-left corner of the DCT array
3. *Feature selection*: assigning the parameters
   N is the dimension of search space/number of host nest
   G is the maximum generation
   C is the total number of cuckoos
4. Generation Step: t
5. While ($t < G$)
   {
         For ($i = 0$; $i <= C$; $i{+}{+}$)
         {
               • Move cuckoo to the new nest with step size S
               • Calculate fitness given by function F_i

```
If (F_i > F_j)
{
        F_j = F_i;
}
```

}

- A fraction Pa of the worst solution is abandoned and new ones are built
- Rank the solutions as per the fitness
- Find the current best solution (nest)
- Pass the current best solutions to the next generation.

}
6. Pick up the solution (nest) with the maximum egg, that is, maximum fitness
7. This solution is the output of the feature selection procedure
8. Classifier

Euclidean distance is defined as the straight-line distance between the two points

$$D = \sqrt{\sum_{i=1}^{N} (P_i - Q_i)^2}$$

where P_i and Q_i are two coordinates of points in dimension i.

The similarity between the extracted features of query image and features of image from the gallery can be measured by the Euclidean distance. The smallest the distance with the image more likely is the required image.

According to the author [24], cuckoo algorithm gives better optimization results than particle swarm optimization (PSO) and thus faces recognition based on the cuckoo algorithm will be more efficient.

Recent studies have demonstrated that CS can outperform other algorithms in many applications [14,25–27].

6.5 Summary and Concluding Remarks

Meta-heuristic algorithms have a wide range of applications in the domains of electrical engineering, computer engineering, business optimization, and many more. CS has also found a prominent importance for the same. Nonlinear optimization problems can be efficiently dealt with CS algorithm. In addition, global convergence of CS can be very fast. However, the forthcoming literature of CS may involve dealing with certain challenges that still need to be resolved.

One such key issue is that there is a need to overcome the considerable gap that exists between practice and theory. Although the algorithms' various applications are in practice, still the algorithms also require mathematical derivation or analysis [28]. In fact, many algorithms do not have theoretical analysis; they just have derived few results about the convergence, improving the convergence, and steadiness about algorithms. Therefore, apart from our knowledge about their working efficiency in practice, it is still difficult to answer the questions such as how we can improve the algorithms and why it works by getting into depth of their procedure of working [29].

Another vital concern is to deal with parameter tuning. As all the algorithms are associated with the parameters, these parameters will impact the overall performance of an algorithm to a great extent. Therefore, all the algorithms require parameter tuning, which itself becomes a problem of optimization. In fact, parameter tuning has emerged as a recent research area [30,31], which requires further studies.

In addition, even the algorithm has found a prominent role in many applications of various discipline, the major concern for the applications is less number of design variables, typically few hundreds. Instead, if the number of variables can be increased to thousands or even more than these then only real-world applications can be optimized with improved results.

All these challenging issues inspire more research in the near future. In addition, with no hesitation it can be concluded that in the forthcoming years we are going to see many more applications of CS, which will further enhance the expanding literature.

Solved Examples

1. What is the physical significance of *step-size parameter* in CS algorithm?

Solution 1

The step size controls the contribution of the levy flight in the solution update phase. A high value of step size would mean that the solution in $(i + 1)$th iteration is primarily governed by the levy flight distance (modeled as step length). As a result, a greater emphasis is placed on exploration. On the other hand, a small value of step size would mean that the levy flight has a little say in the solution update phase; this essentially favors exploitation. Thus, in conclusion, step size can bring about a holistic trade-off between exploration and exploitation to ensure that the algorithm converges to an optimal solution.

CAVEAT: *Do not confuse step size with step length.*

2. Given that the step size α is bounded in the closed interval [0.00,1.50] and maximum number of iterations allowed $N_{max} = 5000$, what is the expected value of step-size parameter after 2000 iterations?

Solution 2

Given,

Maximum value of step size $(\alpha_{max}) = 1.50$

Minimum value of step size $(\alpha_{min}) = 0.00$

Maximum number of iterations allowed $(N_{max}) = 5000$

Iteration after which the value of step size (N) is desired $= 2000$

The expected value of α is

$$\alpha = \frac{e^{\frac{\ln(\alpha_{min})-\ln(\alpha_{max})N}{N_{max}}}}{\sqrt[4]{N}}$$

Substituting all the values, we get

$$\alpha = 0.135$$

3. In what ways does CS algorithm outperform other swarm intelligence algorithms?

Solution 3

The reason for better performance of CS algorithm is a fine balance that exists between randomization and intensification along with less number of control parameters. The balance between randomization (exploration) and intensification (exploitation) is a prerequisite to any swarm intelligence algorithm and in this case it is aptly provided. To add to this, less number of control parameters removes the complexity from the algorithm, whereas at the same time making it more generic.

Exercises

1. Explain the reproduction technique of cuckoo birds. Also relate, how this can be utilized for solving optimization problems?
2. What is meant by CS algorithm?
3. Compare and contrast CS algorithm with other meta-heuristic algorithms.
4. State the traditional CS algorithm.
5. Explain the levy flight walk. How this concept is implemented in CS algorithm?
6. What is meant by local minima and global minima?
7. By what means crossover is useful in CS algorithm? How we can perform crossover and mutation in CS algorithm?
8. How the term exploration and exploitation are associated with meta-heuristic algorithms?
9. State the important unique features of CS algorithm.
10. Explain multiobjective CS algorithm. What is the use of considering multiple objectives for optimization?
11. List the applications or problems involving multiple objectives for optimization.
12. Enrich the application areas of CS algorithm with examples.
13. State the use of gradient for improving the performance of CS algorithm.

14. What are the important issues or parameters to be considered while designing an optimization problem?
15. Implement the CS algorithm in any programming language of your choice.
16. Discuss how certain variants of CS prove to be advantageous over the primary one in certain applications.

References

1. X. S. Yang and S. Deb, Cuckoo search via levy flights, *Proceedings of the World Congress on Nature & Biologically Inspired Computing (NABIC'09)*, IEEE, Coimbatore, India, December 2009, pp. 210–214.
2. A. M. Reynolds and M. A. Frye, Free-flight odor tracking in Drosophila is consistent with an optimal intermittent scale-free search, *PLoS One*, 2007, 2, e354.
3. S. Walton, O. Hassan, K. Morgan et al., Modified cuckoo search: A new gradient free optimization algorithm, *Chaos Solitons & Fractals*, 2011, 44(9), 710–718.
4. X.-S. Yang, Cuckoo search and firefly algorithm, *Studies in Computational Intelligence*, 2014, 516, 1–26.
5. Z. Zhang and Y. Chen, An improved cuckoo search algorithm with adaptive method. In *Computational Sciences and Optimization (CSO), 2014 Seventh International Joint Conference on* 2014, July 4, pp. 204–207. IEEE.
6. K. Chandrasekaran and S. P. Simon, Multi-objective scheduling problem: Hybrid approach using fuzzy assisted cuckoo search algorithm, *Swarm and Evolutionary Computation*, 2012, 5(1), 1–16.
7. K. Choudhary and G. N. Purohit, A new testing approach using cuckoo search to achieve multi-objective genetic algorithm, *Journal of Computational*, 2011, 3(4), 117–119.
8. X. S. Yang and S. Deb, Multiobjective cuckoo search for design optimization, *Computers and Operational Research*, 2013, 40, 1616–1624.
9. S. E. K. Fateen and A. B. Petriciolet, Gradient based cuckoo search for global optimization, *Mathematical Problems in Engineering*, 2014, 2014, 12.
10. R. A. Vazquez, Training spiking neural models using cuckoo search algorithm, *IEEE Congress on Evolutionary Computation (CEC'11)*, pp. 679–686.
11. A. Kaveh and T. Bakhshpoori, An efficient optimization procedure based on cuckoo search algorithm for practical design of steel structures, *International Journal of Optimization in Civil Engineering*, 2012, 2(1), 1–14.
12. A. Kumar and S. Chakarverty, Design optimization for reliable embedded system using Cuckoo search, *Proceedings of 3rd International Conference on Electronics Computer Technology (ICECT)*, IEEE, 2011, pp. 264–268.
13. V. R. Chifu et al., Cuckoo search algorithm for clustering food offers, *IEEE Congress on Evolutionary Computation*, 2014, pp. 17–22.
14. A. R. Yildiz, Cuckoo search algorithm for the selection of optimal machine parameters in milling operations, *International Journal of Advanced Manufacturing Technology*, 2012, doi:10.1007/s00170-012-4013-7.
15. M. Dhivya, and M. Sundarambal, Cuckoo search for data gathering in wireless sensor networks, *International Journal of Mobile Communications*, 2011, 9(6), 642–656.
16. M. Dhivya, M. Sundarambal and L. N. Anand, Energy efficient computation of data fusion in wireless sensor networks using cuckoo based particle approach (CBPA), *International Journal of Communications, Network and System Sciences*, 2011, 4, 249–255.

17. A. Layeb, A novel quantum-inspired cuckoo search for Knapsack problems, *International Journal of Bio-inspired Computing*, 2011, 3(5), 297–305.
18. L. H. Tein and R. Ramli, Recent advancements of nurse scheduling models and a potential path, *Proceedings of 6th IMT-GT Conference on Mathematics, Statistics and its Applications (IC-MSA)*, Universiti Tunku Abdul Rahman: Kuala Lumpur 2010, pp. 395–409.
19. E. R. Speed, Evolving a Mario agent using cuckoo search and softmax heuristics, *Games Innovations Conference (ICE-GIC)*, IEEE, 2010, pp. 1–7.
20. X. S. Yang, *Nature-Inspired Metaheuristic Algorithms*, Luniver Press, Bristol, England, 2008.
21. A. R. Yildiz, A novel hybrid immune algorithm for global optimization in design and manufacturing, *Robotics and Computer Integrated Manufacturing*, 2009, 5, 261–270.
22. N. Baskar, P. Asokan, R. Saravanan and G. Prabhaharan, Optimization of machining parameters for milling operations using nonconventional methods, *International Journal of Advanced Manufacturing Technology*, 2009, 25(11–12): 1078–1088.
23. X.-S. Yang, S. Deb et al., Cuckoo search for business optimization applications, *National Conference on Computing and Communication Systems (NCCCS)*, West Bengal, India, 2012.
24. V. Tiwari et al., Face recognition based on cuckoo search algorithm, *Indian Journal of Computer Science and Engineering (IJCSE)*, 2012, 3, 401–405.
25. H. Q. Zheng and Y. Zhou, A novel cuckoo search optimization algorithm based on Gauss distribution, *Journal of Computational Information Systems*, 2012, 8(10), 4193–4200.
26. A. H. Gandomi, X. S. Yang and A. H. Alavi, Cuckoo search algorithm: A metaheuristic approach to solve structural optimization problems, *Engineering with Computers*, 2013, 29(1), 17–35. doi:10.1007/s00366-0110241-y.
27. A. Noghrehabadi, M. Ghalambaz and A. Vosough, A hybrid power series—Cuckoo search optimization algorithm to electrostatic deflection of micro fixed-fixed actuators, *International Journal of Multidisciplinary Sciences and Engineering*, 2011, 2(4), 22–26.
28. P. Civicioglu and E. Besdok, A conception comparison of the cuckoo search, particle swarm optimization, differential evolution and artificial bee colony algorithms, *Artificial Intelligence Review*, 2011. doi:10.1007/s10462-011-92760.
29. X.-S. Yang and S. Deb, Cuckoo search: Recent advances and applications, *Neural Computing and Applications*, 2014, 24(1), 169–174.
30. A. E. Eiben and S. K. Smit, Parameter tuning for configuring and analyzing evolutionary algorithms, *Swarm and Evolutionary Computation*, 2011, 1, 19–31.
31. V. Bhargava, S. E. K. Fateen, A. Bonilla-Petriciolet, Cuckoo search: A new nature-inspired optimization method for phase equilibrium calculations, *Fluid Phase Equilibria*, 2013, 337, 191–200.

7

Artificial Bee Colony

7.1 Introduction

The field of swarm intelligence has gained popularity as a research interest of scientists belonging to various diverse fields over the recent year. Bonabeau [1] explained swarm intelligence as "any attempt to design algorithms or distributed problem-solving devices inspired by the collective behavior of social insect colonies and other animal societies." His main focus was the behavior shown by social insects such as bees, wasps, termites, and other ant varieties. Artificial bee colony (ABC) is considered a pioneer among the swarm intelligence algorithms, based on the honey bee foraging behavior. This chapter explains the concepts and applications related to ABC in detail.

7.2 Biological Inspiration

Swarm intelligence algorithms works on two fundamental concepts, which are necessary and sufficient for exhibiting their properties, self-organization, and division of labor. The distributed problem-solving system shows swarm intelligent behavior by self-organization and adaptation to their environment. Self-organization is known as a collection of dynamic mechanisms resulting in structures that are formed at the global level with means from interactions with their low-level components. The interaction between the system components is controlled by the basic rules defined in the algorithm. The rules are responsible for ensuring that the interactions in the system are governed by the local information and do not have any relation with the global pattern.

Self-organization properties can be explained as follows:

1. A common behavioral *rule of thumb* is that positive feedback promotes the development of convenient structures. Positive feedback is done by following the recruitment and reinforcement behavior such as dancing of bees, following by ant species, and so on.

2. The positive feedback is needed to be counterbalanced using a negative feedback in order for the collective pattern to be stable. Thus, to prevent saturation that occurs due to various possibilities such as food source exhaustion, number of active foragers, competition, and crowding among food source, the negative feedback mechanism is very beneficial.

3. The fluctuations in the behavior such as errors, random walks, and task switch among the swarm population randomly are very important properties required for innovation and creativity of solutions. Randomness is a crucial property among new structures because it enables formation of new solutions.

Hence, it can be said that self-organization requires that the density of mutually tolerant population should be minimal allowing them to use others' as well as their own results from their activities.

The simultaneous working of different tasks happens inside a swarm by specialized individuals. This concept is referred as a division of labor. It is believed that this simultaneous task performance using the cooperation among unspecialized individuals is more efficient than performing the task sequentially by the unspecialized individuals. The changes in the conditions of search space can also be identified using the division of labor. The two properties, that is, self-organization and division of labor are hence considered necessary and sufficient conditions for the collective performance of a swarm to obtain the swarm intelligent behavior.

7.3 Swarm Behavior

The forage selection's minimal model leading to the development of the overall intelligence of honey bee swarm has essentially three components: (1) food sources, (2) employed foragers, and (3) unemployed foragers. There are also two leading nodes of the behavior: (1) nectar source recruitment and (2) source abandonment.

a. *Food sources*: Food source value depends on factors such as its richness, concentration of energy, proximity to the nest, or ease of extraction of energy. In addition, the *profitability* of the food can be represented to be a single quantity for simplicity.

b. *Employed foragers*: Employed foragers are associated with a food source and they are tasked to exploit that source. They transmit with them the information about their employed source, its position from the nest, and the profitability of that source, and then share this information with others with some probability.

c. *Unemployed foragers*: Unemployed foragers look out for food sources to exploit continuously. There are two types of forager bees: (1) scouts and (2) onlookers. Scout bees search their environment continuously in order to find new food sources, whereas onlookers wait at the nest establishing food sources from the information that is transmitted by employed foragers. The average number of scouts over various conditions comes to 5%–10%.

The information exchange among bees results in the important phenomena of collective knowledge. Some parts of the hive can be observed to be common for the entire hive. Among these parts, the most important area of the hive from the perspective of information exchange can be seen as the dancing area. Dancing area is the place where the bees communicate among them about the quality of the food sources. The dance is referred as waggle dance.

The dance floor provides the information about all the rich food sources to the onlookers, where they can watch numerous dances and then decide which among them would be the most profitable food source. The most profitable sources have a high probability of getting selected by the onlookers because more information is circulated about these food sources during the dance.

The information among the employed foragers is shared with a probability, which directly depends on the profit factor of the food source, and more profit results in more sharing of its information during the waggle dance. Hence, the recruitment of the food source is also directly proportional to the food source's profitability. The key processes involved in collective intelligence by bees are as follows:

Positive feedback: The number of onlookers visit increases with the nectar amount in the food sources.

Negative feedback: When the food source quality becomes poor, the bees stop their exploitation.

Fluctuations: New food sources are continuously discovered by scout bees using the random search process.

Multiple interactions: In the dance area, the bees share information about their food sources with the other bees.

When a food source is located by a bee, the bee uses its capability to memorize the position of the source and then starts exploiting it immediately, changing its status to an *employed forager*. The foraging bee then collects nectar from the source, returns to the hives, and then deposits the nectar to the food store (Figure 7.1) [2]. When the food is deposited in the food store, the bee starts one of the following tasks:

1. It leaves the food source and then change into an uncommitted follower (UF).
2. It starts waggle dance, recruit nest mates and return the same food source (EF1).
3. It returns to the food source for foraging without recruiting other nest mates (EF2).

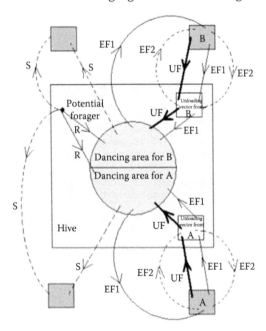

FIGURE 7.1
Different modes in which a bee functions.

7.3.1 ABC Algorithm

The algorithm can be described through the following steps (Figure 7.2):

1. *Initialize*: Define the bee colony as $X = \{x_i \mid i = 1, 2, ..., n\}$, here n is the population size and x_i the ith bee.

2. Calculate the fitness f_i of each employed bee x_i using the fitness function. For each bee, x_i, store the maximum amount of nectar and the corresponding food source.

3. For each employed bee, generate a new solution v_i in the previous solution's neighborhood in its memory by $v_i = x_i + (x_i - x_k) \times \Phi$. Here k is an integer near to i, $k \neq i$, and Φ is a random number in the real domain and between $[-1, 1]$.

4. Update x_i using the greedy criteria. Calculate the fitness of v_i. If it is found that the new solution v_i is better than x_i, then x_i is replaced with v_i; else x_i is left untouched.

5. Calculate the probability value of P_i using the fitness f_i of x_i.

6. The onlookers then use the probability P_i to choose food sources, searching neighborhood for generating candidate solutions, and for calculating their fitness.

7. Update food sources using the greedy approach.

8. Memorize the food source and its nectar amount, which shows the best fitness value.

9. Analyze the solutions and check for abandoned solutions in the search space. If an abandoned solution is found, then replace the solutions with the newly generated random solutions by $x_i = x_{min} + (x_{max} - x_{min}) \times \Phi$, where Φ is a random number in real space between $[0, 1]$, min and max are lower and upper bounds of the solution space, respectively.

10. Terminate the algorithm if the maximum iteration count (k_{max}) is reached or the stop conditions occur. Otherwise repeat steps (3)–(9).

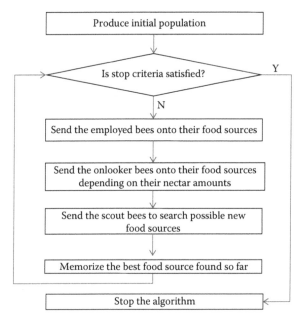

FIGURE 7.2
Flowchart of an artificial bee colony algorithm.

7.4 Various Stages of Artificial Bee Colony Algorithm

Various stages defined in the ABC algorithm are explained as follows:

1. *Initialization of parameters*: The basic ABC algorithm sets three control parameters, that is, the count of food sources (SN), the number of iterations through which further improvement in food source is not observed, and then it is assumed that the food source is to be abandoned from the population, and a termination criterion is set.

2. *Initialization of the population*: The initial population in basic ABC algorithm is set through a random approach. We define $V_i = \{v_{i1}, v_{i2}, ...; v_{in}\}$ as the ith food source and n be the problem dimension. The food source generation can be defined through the following equation:

$$v_i^j = v_{min}^j + \text{rand}[0,1]\left(v_{max}^j - v_{min}^j\right) \tag{7.1}$$

where v_{max}^j and v_{min}^j define the upper and lower bound for dimension j, respectively.

3. *Local search operator*: It is used by both the onlooker bees and the employed bees. The equation for local search operator is as follows:

$$v_{new}^j = v_i^j + \text{rand}[-1,1]\left(v_i^j - v_k^j\right) \tag{7.2}$$

Here, $j \in \{1,2, ..., D\}$, $k \in \{1,2, ...SN\}$ and $k \neq i$ are indexes chosen randomly. When the new solution is obtained, V_{new} is calculated and compared to V_i. The solution with the highest fitness value is then kept while the other is discarded.

4. *Global search operator*: The onlookers in ABC perform exploration search by utilizing a global search operator. Equation 7.3 calculates the probability value that the onlookers use to determine promising areas in the search space.

$$P_i = \frac{\text{fit}_i}{\displaystyle\sum_{i=1}^{n} \text{fit}_i} \tag{7.3}$$

$$\text{fit}_i = \begin{cases} \dfrac{1}{1+f_i} & \text{if } f_i \geq 0 \\ 1 + \text{abs}\left(f_i\right) & \text{if } f_i < 0 \end{cases} \tag{7.4}$$

where fit_i denotes the fitness of the ith solution and SN denotes the number of food sources.

5. *Random search operator*: A food source is termed to become abandoned if its fitness value does not improve after a certain number of iterations. After the food source is abandoned, a new food source is generated randomly and replaces the abandoned one. The basic ABC algorithm assigns at the most one scout bee to a random food source after each cycle.

7.5 Related Work

ABC algorithm has been applied in a variety of problems of the real-world domain. The property of ABC to obtain quasi-optimal solutions in acceptable time has made it a preferred choice for many researchers. For the optimal power flow problem, [3] uses ABC algorithm as its optimizer for adjusting their power system controls optimally. Both the continuous as well as discrete variables are assigned to the control parameters. The paper used different objective functions such as total active power loss, fuel costs (convex and nonconvex), voltage stability enhancement, voltage profile improvement, and emission cost are defined and then the constrained nonlinear optimization problem is solved. The obtained solution's effectiveness and validity is then compared to the IEEE 9-bus, 30-bus, and 57-bus system. The results obtained are then tested and compared with the results of the other meta-heuristic solutions in the literature. After the simulation, the results of ABC algorithm were accurate for different objective functions.

Different objective functions such as total active power loss, fuel costs (convex and nonconvex), voltage profile improvement, voltage stability enhancement, and emission cost are shown effective using the IEEE 9-bus and 30-bus system (Figure 7.3) [3].

In [4], the authors used ABC algorithm to enhance the estimation performance for the fault sections in the power systems. The algorithm mimics the honeybee's foraging behavior, to give exploration and exploitation methods in order to have near-optimal search mechanism. The algorithm excelled in giving decision on external parameters such as mutation and crossover rates, which results in the computation performance improvement. In addition, the method added a new random selection method in order to generate new sources reducing the probability for being trapped in a local minimum, and thus it reduces the faulted sections in the solution space efficiently. This makes the engineers to find precise fault sections in the alarm systems that reduce the chance for wrong diagnosis.

In [5], ABC algorithm is employed to determine the protein structure computationally and experimentally. The experimental methods of the literature are usually considered expensive and time-consuming and are not feasible for protein structure identification. So, the problem is solved by computational methods by converting it into an optimization problem in which the objective is to find the lowest free energy. The ABC algorithm is

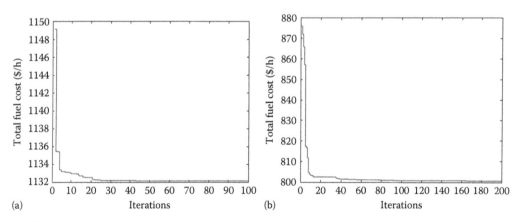

FIGURE 7.3
Graph showing the variation of total fuel cost with each iteration of the algorithm.

adapted for this problem to search the protein conformational search space and the lowest free energy conformation is found.

A variant of ABC algorithm using the distributed approach was proposed in [6]. In this parallel algorithm, several subgroups of the bee colony are created. A local search is then performed in each subgroup concurrently on different processor nodes. The nodes then exchange their local best solutions among each other. The message passing technique is utilized by this algorithm. The experimental results for the algorithm show better solutions in terms of quality and computing time on comparison with the sequential ABC.

Recently, there have been significant advances pertaining to the use of nature inspired computing algorithms in control engineering. The primary use of such algorithms is to figure out what parameters need to be selected and how the parameter values can be fine tuned to get optimal performance. Upgraded artificial bee colony (UABC), a variant of the original ABC algorithm, does find its utility in this field as a result of rigorous tests performed on standard benchmark functions [7]. The results found were quite satisfactory and in some instances they even outperformed the Karaboga algorithm.

In [8], two variants of ABC are proposed that use new ways of position updation for the artificial bees. Various empirical tests were conducted to compare these variants with the original ABC and several other meta-heuristics present in the literature. The experimental results observed inferred that the ideal values of parameters depended on the hardness of the optimization problem and the values, which are specified in the existing literature, are to be used carefully. The usage of onlooker bees (OBs) in the optimization problems is also considered advantageous in some situations, whereas disadvantageous in others. In addition, a potential setback of ABC was observed in which it obtained less optimal results when the solution is located at the centre of the search space.

A recently introduced paradigm of algorithms is that of Memetic computation (MC) [9]. They efficiently solve complex optimization problems. ABC algorithm is also known to generate a good solution for complex combinatorial and continuous problems. A greater performance can be obtained by combining both these concepts into one. This results in the development of a new Memetic ABC (MABC) algorithm. MABC algorithm is developed in hybridization with two different local search algorithms: (1) the random walk with direction exploitation (RWDE) and (2) the Nelder–Mead algorithm (NMA). The former is more inclined toward exploitation of solution space, whereas the latter toward the exploration of the solution space. Thus, to maintain a balance between exploration and exploitation, a stochastic adaptation rule is used. This MABC algorithm was applied to large-scale continuous global optimization and gave encouraging results [9].

The ABC is shown to exhibit insufficiency when it comes to its solution search equation, which shows good exploration properties but poor exploitation properties. The authors in [10] employed a modified method of ABC known as improved ABC with modified search equation (CABC); the candidate solution is generated using a modified search equation, so as the solution search ability is improved from the conventional ABC. In addition, an orthogonal experimental design (OED) strategy is used in order to have an ohogonal learning (OL) algorithm, so as to find more useful data in the search experiences. It is known that OED's perform good sampling of small number of representative combinations that are required for testing and thus constructs more efficient candidate solutions.

In [11], the efficiency of ABC is compared with improved bees algorithm and harmony search algorithm on some benchmark unimodal and multimodal optimization problems. The main objective of this comparison was to check the effect of dimensionality of the problem on the algorithm's performance and the control parameter's effect on the algorithms.

The authors in [12] applied ABC algorithm with fuzzy clustering in order to classify datasets of cancer, heart disease, and diabetes from the UCL database. The clustering was done, so as to identify the different multidimensional medical data from the data based on their similarity factors. Similarity factors for the data points were identified using the distance measurement. In the fuzzy clustering approach, the members in the unique clusters had a fuzzy membership value between 0 and 1, which identified its degree of membership to the clusters.

The process of symbolic regression [13] deals with defining a mathematical model with only a finite sampling of the independent variables and their respective values of the dependent variables. The authors solved a set of benchmark problems relating to symbolic regression using the ABC algorithm and its results were compared to those of genetic algorithms (GAs) and other evolutionary algorithms. The simulations showed promising results for the ABC algorithm in terms of robust and feasible solution values.

In the case of binary structured problems, the original ABC cannot be directly applied for optimization. Therefore, in [14] a new modification for the original ABC was introduced, designed particularly for binary optimization. It applied a new differential expression, which used a measure of dissimilarity among the binary vectors rather than the vector subtraction, which was used in the original ABC algorithm. This modification helped the algorithm to show primary characteristics of the original ABC algorithm, which can be applied for the binary optimization problems. The DisABC's differential expression was able to be applied to the continuous space such as the original ABC and its result is used for a two-phase heuristic in order to define a solution in the binary space. The DisABC algorithm was then checked for solving uncapacitated facility location problem (UFLP).

Some modifications on the original ABC iteration equation were done inspired by the strategies employed by the PSO algorithm in [15]. The first item is appended with inertial weight in order to balance the global and local searching process. The second item is introduced with a contractive parameter rather than a random value, showing the nonlinear descending property and has a contractive effect on the algorithm's search space. In addition, a random disturbance is introduced to the ABC's renewal stage, which increases the algorithm's efficiency in the later stages of the iteration and increases the accuracy of the results. The proposed algorithm was then tested for a benchmark optimization function and then it was applied to a data clustering problem for the analysis of gene data on the DNA set.

[16] presented a hybrid Pareto-based discrete ABC algorithm, which was then applied to solve the multiobjective flexible job shop scheduling problem. The solution in the hybrid approach was defined as a food source, composed of two parts: (1) the routing part and (2) scheduling part. Each part is to be assigned with discrete values. The employed bees exchange valuable information from each other by means of a crossover operator. The nondominated solutions are recorded and stored in an external Pareto archive set. The algorithm introduced a fast Pareto set update function. The exploration and exploitation abilities of the algorithm were balanced using several local search approaches. The effectiveness of the algorithm was checked by experimenting on several benchmark problems and its comparison with the recently published algorithm was done.

The framework of the pareto-based ABC (PABC) can be summarized as follows:

1. Initialization phase
    a. Set the system parameters
    b. Initialize the population

2. Use the Pareto nondominated sorting function over the population, and then change the Pareto archive set by using the first Pareto level front solutions.

3. Output nondominated solutions of the Pareto archive set if the stopping conditions are met, otherwise move to Step 4.

4. Employed bee phase.

   a. Assign an employee bee to each population's solution.

   b. Perform local search for each employee bee on the respective solution and then form a new neighboring solution.

   c. Assign the better solution among the new ones to be put in the population and record the old and the new solution. In case of two nondominated solutions, select any one among them randomly.

   d. If a solution has not been improved through limit cycles, then the corresponding employed bee becomes a scout bee and performs Step 6.

   e. Employed bees perform crossover among each other, so as to learn from others.

   f. Calculate the solutions corresponding to the employed bees and then use Pareto nondominated sorting on the newly obtained population. Then modify the Pareto archive set through the first Pareto level solutions.

5. Onlooker bee phase

   a. Every onlooker bee chooses three solutions randomly from the population. It then selects the best among the three. If there is no dominant solution among the three, randomly choose one of the three solutions.

   b. Every onlooker bee then uses local search strategy to select a food source and then records the better solutions of the population using greedy strategy.

   c. The solution with respect to each onlooker bee is calculated. Use the Pareto non-dominated sorting over the newly obtained population. Then modify the Pareto archive set through the first Pareto level solutions.

6. Scout bee phase

   a. A random food source is selected by the scout bees and several local searches are applied in the region. Perform greedy selection procedure on the newly generated solution.

   b. The solution with respect to each scout bee is calculated. Use the Pareto non-dominated sorting over the newly obtained population. Then modify the Pareto archive set through the first Pareto level solutions.

7. Return to step 3.

An improvement of ABC algorithm in the scout bees' phase was proposed in [17]. Logistic equation is employed to generate the initial population, so as to improve the global convergence. The results of the algorithm were tested for six benchmark function and good performance was observed for the improved ABC than other modified algorithms for the optimization problems. Initialization of population affects greatly the quality of final solutions in the evolutionary algorithm as it decides greatly the convergence speed and prevents premature stopping. The chaotic motion is known to show randomness, ergodicity, regularity, and so on; it is known to iterate over the whole space and thus improves the search ability of the algorithm. Therefore, in order to initialize the population, the algorithm employs a chaotic map, which increases the convergence speed and the diversity of the solution. Thus, the new chaotic system employed in the algorithm prevents the premature convergence of the ABC algorithm. A chaotic sequence is generated using piecewise logistic equation showing more randomness. The initial solutions in this approach are generated through piecewise logistic chaotic mapping, in which the randomness of the population is maintained and the ergodicity of the solution is improved. The basic algorithm then uses Equation 7.1 to jump out of the local optima by searching the global scope. If Equation 7.1 is directly used by the bees, then less convergence efficiency is observed by the algorithm.

[18] proposed a fast synthetic aperture radar (SAR) image segmentation procedure, which was based on ABC algorithm. The algorithm defines threshold estimation as a search procedure, which finds appropriate value in a continuous grayscale interval. So, optimal threshold is obtained through ABC algorithm. The original number gets decomposed by the discrete wavelet transform and an effective fitness function is obtained by defining gray number in the gray theory. The approximation image is then applied to the noise reduction, so as to get a filtered image with low-frequency coefficients. High-frequency coefficients are used to reconstruct the gradient image. The filtered and the gradient image-based cooccurrence matrices are thus constructed, and the fitness function of the ABC algorithm is defined using two-dimensional gray entropy. Finally, the optimal threshold is defined using intelligence of the onlooker, scout, and employed bees. Experimental results showed better performance than GA and artificial fish swarm algorithm. The working of SAR image segmentation is represented by Figure 7.4 [18].

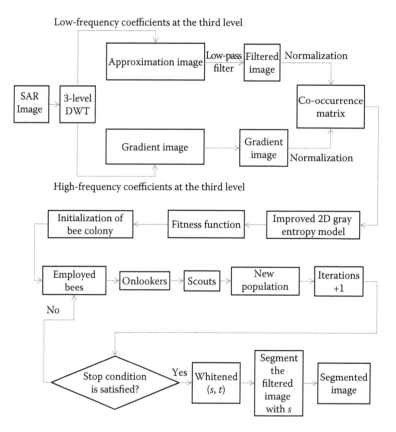

FIGURE 7.4
Block diagram of SAR segmentation method.

The authors in [19] proposed a parallel version of the ABC algorithm employed in shared memory architectures. The bee colony is clustered equally among the processors that are available in the system. The local memory of each processor stores the set of solutions and a copy of every solution is also kept in the global memory. In each iteration, the ABC algorithm improves the solutions at the local memory. After the end of the iteration, the solutions get distributed to their slots in the shared memory, which is then made available to the bees. This approach increases the performance of the algorithm to a great extent.

A new modified algorithm is proposed in [20]. The ABC algorithm maintains a balance between the local exploitation with the global exploration. In the algorithm, initial solutions are first generated using multiple strategies giving importance to the diversity and quality of the food sources. After the initial solution is generated, crossover and mutation strategies are employed to generate a well-defined operation sequence of food sources to be acquired by the employed bees. In addition, a local search strategy is employed based on the critical path and is embedded to the search mechanism in order to enhance the searching behavior and local intensification of the onlooker bees. An initialization strategy for scout bees is also proposed to enhance the updating mechanism, which also results in an enriched searching behavior and avoids premature convergence. In addition, a left shift decoding is used, which transforms solutions in active schedules. This algorithm is then tested on various benchmark functions and is compared with other existing algorithms. The numerical simulations show effective results on this algorithm. The flowchart of the modified algorithm is shown in Figure 7.5 [20].

A hybrid artificial bee colony (HABC) is proposed in [21] that introduces crossover operator of genetic algorithm to ABC and is employed for solving the data clustering problem. An incentive mechanism is employed in HABC to mitigate social learning with the information exchange among the bees by using the crossover strategies commonly applied in genetic algorithm to ABC. Ten benchmark functions were applied in order to assess the performance of the new approach, and it was observed that HABC showed signification improvement from the original ABC algorithm along with several meta-heuristic approaches. The data-clustering problem is then tried to solve using HABC with six datasets from the UCI repository. The results obtained were better than many of the standard algorithm as shown in Figure 7.6 and established the HABC as a competitive algorithm.

Working of HABC is defined below:

1. From the current food sources, choose a group of parent population according to the food sources fitness. The food sources and the parent of the population occupy equal quantity in the space. The sources having high fitness show greater probability of being selected, making sure that the next offspring have all the good characteristics from their parents, and thus are good.

2. For each food source si in original population P_0, select two parent food sources randomly from P_p and cross them. The newly produced offspring is then compared with the si, a greedy-based selection is used and the better solution is retained in the solution space.

The parent population is chosen by the parent selection procedure of the binary tournament. Two food sources are first randomly chosen from the solution set and then the one having the higher fitness is selected as the parent population. This process continues till the required quantity of parent population is reached [21].

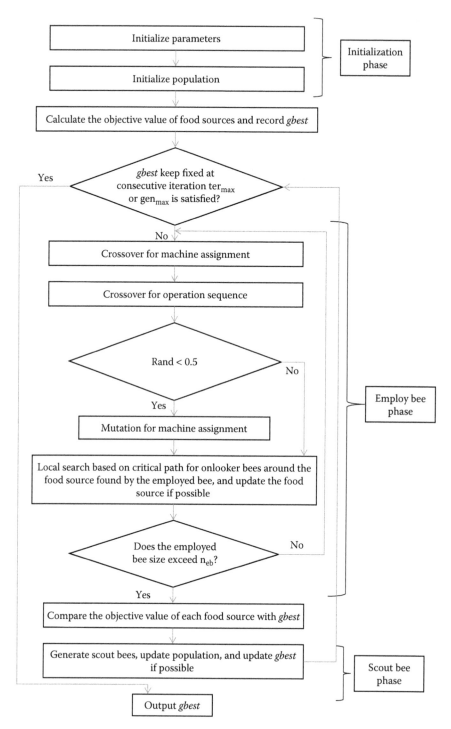

FIGURE 7.5
Flowchart showing different roles of the bees in the modified algorithm.

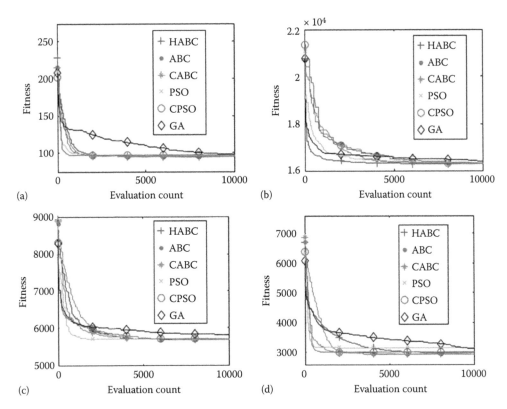

FIGURE 7.6
Graphs showing performance in terms of mean minimum total within cluster variance of algorithms on different (a) Iris data, (b) wine data, (c) contraceptive method choice, (d) Wisconsin breast cancer bench mark functions.

The ABC algorithm is applied on the redundancy allocation problem (RAP) in [22]. This problem sets reliability objectives for its subsystems, so as to meet the constraint for resource consumption such as the total cost. RAP has been considered a promising research area for many decades. The maintenance of feasible results with respect to factors such as cost, weight, volume, and other nonlinear constraints are considered quite difficult in RAP. The authors investigated the nonlinear mixed-integer reliability problems in which it is required to decide both the component reliability in each subsystem and the number of redundancy components in order to maximize the reliability of the system. The study on the reliability design is quite old, with solutions being of mathematical programming or heuristic approaches. The ABC algorithm is thus applied to check the quality of the solution. The algorithm searches over the feasible and infeasible regions in the solution space to obtain near-optimal solutions. The simulation showed that the new algorithm performs good on the reliability redundant allocation design problems.

In [23], a substitutive approach is employed using the artificial neural networks (ANNs) based on ABC with a learning process for solving optimization problems and superiority over S-systems is explained. S-systems are defined as approximate nonlinear models based on power-law formalism and they give a framework for effective simulation of integrated biological systems, which exhibit complex dynamics, such as genetic circuits, showing signal transductions and other metabolic networks. S-systems are, however, known

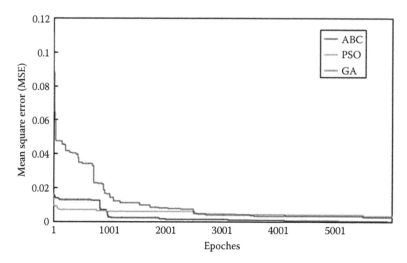

FIGURE 7.7
Figure showing the variation of mean square error with respect to different epochs.

to show a disadvantage in the form of slow convergence, thus requiring large number of iteration for getting convergent profiles of genes. The author thus proposed their approach of combining ANNs with ABC algorithm for the problem. The proposed system effectively generated convergent gene expressions in less iteration than the S-systems and is an effective approach for the problem (Figure 7.7) [23].

Magnetic resonance (MR) brain image classification has also been achieved through the use of ABC algorithm. There have been quite a lot of algorithms, which have been recently proposed to be applied in this area. The authors in [24] proposed a hybrid method of classification using the forward neural network (FNN) optimized through the ABC algorithm. The algorithm was used to classify MR images into normal or abnormal. The algorithm performs in various steps. The first step employs the discrete wavelet transform, so as to extract features from the brain images. In the second step, the feature obtained from the first step is applied to principle component analysis (PCA) to reduce the feature size. In the third step, these reduced features are then fed to the FNN, in which ABC is employed for parameter optimization. The ABC algorithm is modified based on the chaotic theory and fitness scaling and is referred to as scaled chaotic artificial bee colony (SCABC). This process is represented by Figure 7.8.

The algorithm employs a stratified K-fold cross-validation, so as to avoid overfitting in the data. The experiment is then conducted on a real dataset of T-2 weighted MRI images. These images consisted of 66 brain images. The solution is then compared with

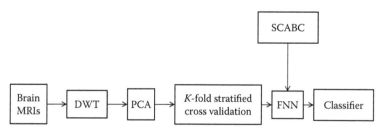

FIGURE 7.8
Block diagram of SCABC method. (Courtesy of The Electromagnetics Academy.)

algorithms in literature such as BP, GA, migration-based elite GA, ABC, and simulated annealing. The proposed algorithm was observed to show 100% classification accuracy and least mean MSE on applying for 20 iterations.

The algorithm consisted of five steps:

1. Extract features using discrete wavelet transform (DWT).
2. Reduce feature size using PCA.
3. Prevent overfitting using stratified *K*-fold cross-validation.
4. Construct the classifier using FNN.
5. Train the FNN using SCABC.

The performance of ABC is considered better than many other approaches such as DE, GA, PSO, and so on. In addition, several algorithms such as FNN are observed to perform better than ABC. It is also worth noting that the performance of algorithm highly depends on the application area they are being applied to [24].

Solved Examples

1. Solve the TSP and QAP problem using the ABC algorithm and tabulate the results using different datasets.

Solution

In order to solve TSP and QAP problem, the ABC algorithm is implemented in C++ on Code Block IDE. Different datasets have been used to solve the QAP problem (Table 7.1).

For the TSP problem, BERLIN52 dataset is used; the following results are obtained (Table 7.1).

Similarly, the results obtained for QAP are provided in Table 7.2.

TABLE 7.1

Results Obtained for TSP Problem When Solved Using ABC Algorithm in BERLIN52 Dataset

Given Optimal	Mean over 5 Runs (500 Iterations Each)	Best Solution	Worst Solution
7500	9820	9049	11,044

TABLE 7.2

Results Obtained for QAP Problem When Solved Using ABC Algorithm

Dataset	Number of Iterations	Given Optimal	Mean over 5 Runs	Best Obtained	Worst Obtained
Nug12	500	578	590	578	608
Esc16i	50	14	14	14	14
Kra32	200	88,700	94,358	91,370	96,580

2. If the maximum and minimum values of food source for a dimension d are 23 and 5, respectively, then show how a food source is initialized in the beginning of the algorithm.

Solution

In order to initialize a food source, we generate a random number in the range [0,1]. In our case, the number obtained is 0.76.

Using Equation 7.1, we have

$$v_i^j = v_{min}^j + \text{rand}[0,1]\left(v_{max}^j - v_{min}^j\right)$$

Here, the values for v_{max} and v_{min} are 23 and 5, respectively. Therefore, putting the values, we have,

$$V_i = 5 + 0.76 * (23 - 5)$$

$$V_i = 18.68$$

Exercises

1. Discuss the significance of waggle dance in the ABC algorithm.

2. What factors need to be taken into account while initializing the population in ABC algorithm?

3. Why is there a need to memorize the best food in an iteration?

4. Contrast the role of onlooker bees with that of scout bees.

5. Discuss how a trade-off is achieved between the exploration and exploitation in the ABC algorithm.

6. What is the consequence of having a positive feedback in the ABC algorithm. Contrast it with the presence of negative feedback?

7. Why only a small fraction of scout bees are present at any time? Discuss in context to their role in the swarm intelligence?

8. Write a short note on the prospects of swarm intelligence in the field of artificial intelligence.

9. Can presence/absence of self-organization affect the convergence of the algorithm? Discuss.

10. ABC algorithm is considered to be a *minimal model of forage selection*. Discuss.

References

1. E. Bonabeau, M. Dorigo and G. Theraulaz, *Swarm intelligence: from natural to artificial systems.* Oxford university press, 1999.
2. D. Karaboga and B. Akay, A comparative study of artificial bee colony algorithm. *Applied mathematics and computation,* 2009, 214(1), 108–32.
3. M. R. Adaryani and A. Karami, Artificial bee colony algorithm for solving multi-objective optimal power flow problem, *International Journal of Electrical Power and Energy Systems,* 2013, 53, 219–230.
4. S. Huang and X. Z. Liu, Application of artificial bee colony-based optimization for fault section estimation in power systems, *International Journal of Electrical Power and Energy Systems* 2013, 44, 210–218.
5. H. A. Bahamish, R. Abdullah and R. A. Salam, Protein tertiary structure prediction using artificial bee colony algorithm, *Third Asia International Conference on Modelling & Simulation,* IEEE, 2009, 258–263.
6. A. Banharnsakun, T. Achalakul and B. Sirinaovakul, Artificial bee colony algorithm on distributed environments, *Second World Congress on Nature and Biologically Inspired Computing,* IEEE, December 15–17, 2010, pp. 13–18.
7. I. Brajevic and M. Tuba, An upgraded artificial bee colony (ABC) algorithm for constrained optimization problems, *Journal of Intelligent Manufacturing,* 2013, 25, 729–740.
8. K. Diwold, A. Aderhold, A. Scheidler and M. Middendorf, Performance evaluation of artificial bee colony optimization and new selection schemes, *Memetic Computing,* 2011, 3(3), 149–162.
9. I. Fister, I. Jr. Fister, J. Brest and V. Žumer, Memetic artificial bee colony algorithm for large-scale global optimization, *IEEE World Congress on Computational Intelligence,* 2012.
10. W. Gao, S. Liu and L. Huang, A novel artificial bee colony algorithm based on modified search equation and orthogonal learning, *IEEE Transactions on Cybernetics,* 2013, 43(3), 1011–1024.
11. D. Karaboga and B. Akay, Artificial Bee Colony (ABC), Harmony search and bees algorithms on numerical optimization, In *Innovative production machines and systems virtual conference,* 2009.
12. D. Karaboga and C. Ozturk, Fuzzy clustering with artificial bee colony algorithm, *Scientific Research and Essays,* 2010, 5(14), 1899–1902.
13. D. Karaboga, C. Ozturk, N. Karaboga and B. Gorkemli, Artificial bee colony programming for symbolic regression, *Information Sciences,* 2012, 209, 1–15.
14. M. H. Kashan, N. Nahavandi and A. H. Kashan, DisABCpp: A new artificial bee colony algorithm for binary optimization, *Applied Soft Computing,* 2012, 12, 342–352.
15. X. Lei, X. Huang and A. Zhang, Improved artificial bee colony algorithm and its application in data clustering, In *Bio-Inspired Computing: Theories and Applications (BIC-TA), IEEE Fifth International Conference,* 2010, pp. 514–521.
16. J. Q. Li, Q. K. Pan and K. Z. Gao, Pareto-based discrete artificial bee colony algorithm for multi-objective flexible job shop scheduling problems, *International Journal of Advanced Manufacturing Technology,* 2011, 55, 1159–1169.
17. H. Liu, L. Gao, X. Kong and S. Zheng, An improved artificial bee colony algorithm, *25th Chinese Control and Decision Conference,* IEEE, 2013, pp. 401–404.
18. M. Ma, J. Liang, M. Guo, Y. Fan and Y. Yin, SAR image segmentation based on artificial bee colony algorithm, *Applied Soft Computing,* 2010, 11, 5205–5214.
19. H. Narasimhan, Parallel artificial bee colony (PABC) algorithm, *World Congress on Nature & Biologically Inspired Computing,* IEEE, 2009, pp. 306–311.

20. L. Wang, G. Zhou, Y. Xu, S. Wang and M. Liu, An effective artificial bee colony algorithm for the flexible job-shop scheduling problem, *International Journal of Advanced Manufacturing Technology*, 2012, 60, 303–315.
21. X. Yan, Y. Zhu, W. Zou and L. Wang, A new approach for data clustering using hybrid artificial bee colony algorithm, *Neurocomputing*, 2012, 97, 241–250.
22. W. C. Yeh and T. J. Hsieh, Solving reliability redundancy allocation problems using an artificial bee colony algorithm, *Computers & Operations Research*, 2011, 38, 1465–1473.
23. W. C. Yeh and T. J. Hsieh, Artificial bee colony algorithm-neural networks for S-system models of biochemical networks approximation, *Neural Computing and Applications*, 2012, 21, 365–375.
24. Y. Zhang, L. Wu and S. Wang, Magnetic resonance brain image classification by an improved artificial bee colony algorithm, *Progress in Electromagnetics Research*, 2011, 116, 65–79.

8

Shuffled Frog Leap Algorithm

8.1 Introduction

In this century, technology is ubiquitous. Much of the innovations taking place today are inspired by a need to automate human processes and make life simpler. Nowadays, most of the technological innovations taking place are a direct consequence of advances in soft computing paradigm. Roughly put, soft computing basically refers to utilization of inexact solutions to get the solution of complex problems (such as NP-Hard problems) for which there is no known algorithm in existence. Soft computing differs significantly from conventional (hard) computing techniques in the fact that it is tolerant to imprecision and uncertainty that can occur in the problem or its solution. Generally, soft computing algorithms are nature-inspired algorithms in a sense that they somehow mimic the natural processes. Phenomena such as behavior of bees, movement of ants, and so on, can be very effectively modeled by using soft computing techniques. The same approaches can also be used to solve similar problems occurring in engineering and management domains.

Soft computing is basically an assemblage of many independent components such as

1. Nature Inspired Algorithms
2. Machine learning
3. Fuzzy logic
4. Evolutionary algorithms
5. Chaos theory

Many nature inspired computing algorithms have been devised to solve complex engineering problems. Each of these algorithms has its own share of pros and cons. As a result, in any practical application, usually a combination of two or more algorithms is used. Such algorithms are called as hybrid algorithms. Among many engineering problems, traveling salesman problem (TSP) is of particular interest. A lot of algorithms have been proposed for solving this complex problem. This chapter discusses the use of shuffled frog leap algorithm (SFLA) to solve this problem. This chapter also sheds some light on finding the solution of the quadratic assignment problem (QAP) with the same algorithm. The efficiency of the algorithm is established through simulation results obtained while solving the aforementioned problems.

8.2 Related Work Done

In [1], the economic dispatch problem has been solved by a recent evolutionary algorithm called the SFLA. The objective of the problem is to get the optimal arrangement of power generations and fuel options given in a fixed load demand while at the same time fulfilling equality as well as inequality constraints and minimizing the total generation cost. In the solution to this problem, a single quadratic function is used to model the thermal-generating unit cost function. In addition, the power balance constraint is considered in addition to the limits of generation capacity. In the approach that is put forward, the optimal combinations of fuel options for the committed generating units can be identified with the application of SFLA. Noniterative Lagrangian multiplier method is used to compute the fitness of each decision vector in the population. The decision vector of SFLA consists of a sequence of integer numbers representing the fuel options available for the generating units. The results of various methods have been compared with the simulation results of the proposed algorithm. The outcome of comparison showed that SFLA algorithm provides quality/optimal solutions in less time.

In [2], unit commitment (UC) problem is dealt with the use of binary shuffled frog leaping (BSFL). UC is the task of determining the optimum schedule of generating units by reducing the overall cost of the power generation while at the same time satisfying a set of system constraints. Researchers have concluded that the standard SFL cannot converge properly and several improvements were suggested in order to cross the barricades associated with the standard SFL. SFL and PSO were merged to form what is called a new memetic algorithm. A cognition component was introduced to improve the stability and global search ability of the original SFL algorithm. This strategy also sees a chaos search being combined with SFLA for better performance. Here, the direction and the length of each frog's jump were extended by mimicking the frog's perception and action uncertainties, which in turn introduced a new frog leaping rule. Zhen [25] also proposed a new leaping rule as well as a new way of dividing the population. The improved SFL algorithm is encoded in the discrete space and a new binary variant of SFL is introduced. As a result, the combiner algorithm is called BSFL. The striking variation between the continuous and binary SFL is that in the binary version, the worst position update entails the switching between "0" and "1" values. This switching needs to be done as per the leaping rule. The motivation is to update the position in such a way that the current bit is changed with a probability. The probability value is computed in advance according to the leaping rule. To improve on the performance of the obtained BSFL, yet another modification is done on BSFL so that the obtained algorithm is improved BSFL (IBSFL). To demonstrate the efficacy of the proposed algorithm, the algorithm is tested on a UC problem. Using the outcomes obtained by the BSFL, IBSFL and the results available in the literature, it was concluded that the IBSFL is a very effective algorithm in giving optimal solutions for UC problem.

[3] proposes a novel artificial intelligence-based method. This method is applied to an e-course composition by incorporating the notion of personalized learning. E-learning refers to learning procedure that often employs computers, electronic devices, and the Internet. Development of such systems helps the instructors in composing e-courses. However, the lack of *personalized learning* in the previous approaches could not justify with the difference in the individual's prior knowledge and his or her gaining capabilities. The

personalized learning strives to provide materials that are tuned to an individual learner. Hence, to verify and validate the proposed algorithm, BSFLA is compared with binary particle swarm optimization (BPSO) and genetic algorithms (GAs). Results confirmed that the proposed algorithm outperforms BPSO, GA, and so on.

[4] proposes the varied version of SFLA. The proposed modified SFLA derives the expression for optimal placement of switches (both static and dynamic) in distribution automation systems (DASs). DASs comprise remote-controlled and automatic switches. The switches play a crucial role in defining the reliability of the system. The key objective is achievement of high-system reliability for the distribution sector with minimization of capital costs. Consideration of multiple objectives is handled with a fuzzy approach; in addition, fuzzy membership function is defined for each term expressing the objective function.

The objective of the problem-solving method is to determine the optimum number, location, and type of manual and automatic switches in DASs. This involves the following:

1. Minimizing the customer interruption cost (CIC) for improving the reliability.
2. Minimization of switched purchasing and maintenance cost (SPMC).

Therefore, the objective of providing electrical energy to customers while at the same time maintaining symmetry between the important elements, such as cost, product support, and reliability, is fulfilled by finding the best solution to the above-mentioned optimization problem. The IEEE 123-node feeder standard test system is used to study the simulated performance of the proposed approach. The simulation results confirm the potential capability of the proposed method. In addition to this, the algorithm is also compared to other optimization methods such as GA to establish its accuracy and efficiency. The results showed that the efficiency gain of the modified SFLA method provides an efficient way to get to the bottom of the discrete optimization problem presented here. To summarize, when the proposed optimization algorithm is used, the sum of customer compensation and switching costs can be considerably reduced with the planning of a new distribution network.

[5] discusses a novel application of SFLA for robot path planning in partially known environments. The proposed algorithm allows for a potential mobile robot to find its path through static obstacles and navigate, so as to reach from its initial position to the target without colliding with other objects. The environment is assumed to be partially unknown as the detection range and accuracy of Khepera's sensor is bounded. In addition, each obstacle is assumed to be a single point placed in center of the obstacle. This position information of an obstacle is known prior to our algorithm. However, the shape, size, and other geometries from the environment are unknown and the robot path planning processor keeps on updating this information on its path. The general procedure for real-time path planning is organized as follows:

1. Transformation of path planning problem into an optimization problem
2. Definition of the optimization objective function
3. Application of SFL to solve the robot's path optimization problem. It is during this process that the position of the globally best frog in each iteration is selected; this phenomenon causes the robot to move to its target in steps

It is worth noting that the path generated by the robot in each step is known to be just feasible, that is, the robot can eventually reach its goal on the circumstance of no collision with obstacles. We cannot assert that the obtained path provided by the algorithm is globally optimal even when the programmed path has some optimal effect on the quality of the solutions generated. Furthermore, the smoothness of the robot's move is demonstrated with the simulation of this new algorithm. The results reveal that these new algorithms are more effective and robust.

In [6], a new frog leaping rule has been put forward. The unique feature of this approach is the use of exponentially decaying uncertainties in order to improve performances of the SFLA. As a result, the SFLA has been modified to widen the local search space that helps the algorithm from falling into local minima at the first few iterations. The narrow local search space then accelerates convergence speed at the later iterations. In this study, the searching performance of the SFLA is improved. Following this, the proposed modified SFLA is used for the tuning of fuzzy controller's parameters. The fuzzy control has emerged as a control technique, which is intelligent because it resembles the human decision-making process using imprecise knowledge and uncertain conditions. The complex systems with a bunch of uncertainties can be dealt with the fuzzy logic. For example, the approach is best suitable for industrial processes. Traditionally, fuzzy controllers were constructed in a heuristic way; application-specific knowledge was used in the construction process. Designers modeled their system knowledge and work experiences about the process in the form of fuzzy rules; these fuzzy rules describes the linguistic relationship that exists between the input and output variables. For this particular application, the fuzzy controller is designed without taking into consideration the expert knowledge. The effectiveness of the SFLA-based fuzzy controller is demonstrated using a ball and beam system.

[7] proposed a new search strategy in order to enhance the efficiency of the original SFLA. The motivation behind this improved algorithm comes from the fact that once some optimizations are run frog's position becomes closer; and this happens in each memeplex. Needless to say, this problem eventually leads the algorithm to fall in local minima or a premature convergence. In order to a better efficiency, a new search strategy has been proposed; this proceeds by infecting not only the worst individuals in the population but also the best individual. To further enhance or accelerate the convergence process, two factors in the search strategy are formulated. Instead of changing the position of only the worst frogs, best frog's position is also modified to perform a local search, so as to enhance the efficiency by improving their performance. In the proposed enhanced algorithm, not only the worst frog but also the best frog changed their position. The best frog implements the same strategy for changing its position as worst frog in the standard algorithm. Although the modified updated position of the best frog is based on the distance between the global best frog P_g and worst frog P_w. As this distance is of consequence, it ensures that there will occur a major leap from P_b (best frog) to P_g (global best frog) if the fitness value of P_b is not improved. The movement of the latter is based on the distance between P_g and its former position. If this process fails to improve the best frog performance, it continues to maintain its current position in the memeplex.

Worst frog assessment becomes more practical by increasing the fitness/quality of the best frog. Thus, individuals will be well distributed in the search space within each memeplex. The improved algorithm can be described as follows:

1. First determine the best and the worst frog. Let P_{bi} and P_{wi} be the frogs with the best and the worst fitness, respectively within the ith memeplex. P_g identifies the global best frog.

2. Improve the best frog's position to increase its fitness. Evaluating and updating the position of the best frog are carried out first.

3. The worst frog's position is updated for improving the convergence of the memeplex.

4. Repeat step 3 for the global best frog. If there is no further improvement possible, proceed to the next step.

5. Generate a new random solution in order to replace the worst frog P_{wi}.

In [8], the authors have presented a novel algorithm that is based on chaos search being implemented over SFLA. The algorithm uses chaos search to produce neighborhoods of extremum in order to sustain a variety of solutions and not to fall into the local optimal solution when the evolution of the individuals stops. Chaos is a general phenomenon present in a nonlinear system and it has got the characteristics of ergodicity, random city, regularity, and sensitivity to the initial conditions. A chaos search has the ability to explore all the points in the search space that too without any repetition. As such, the chaos search algorithm shows its usefulness in global search in which it is used to avoid local optimal. The key logic of implementing composite shuffled frog leaping algorithm (CSFLA) can be put as follows: make SFLA processes as the most important processes and when the global best frog X_g stops evolution, use chaos search in order to generate a new optimal solution in the neighborhood of X_g. This is essentially used to pull out of the local optimal solution.

Finally, the CFSLA is used to solve mid-long term optimal operation problem found in cascade hydropower stations and the results when plotted show its feasibility and high efficiency in comparison to other two algorithms.

Materialized view selection [9] has been significantly acknowledged in the past few decades owing to its variety of application in diverse fields, such as query speedup, data warehouse, update processing, and decision support systems. Materialized views are known to be especially attractive in data warehousing environments (data warehouses project a query intensive nature). Materialized view speedups the query by precomputing the queries that are frequently asked and these queries are stored for future reference. However, apart from the occupying extra disk space, the consistency of the data must be maintained between the data stored on the disk with the one present online. The materialized view selection has proved to be an NP-hard problem and is also essentially a nontrivial task. The purpose of materialized view selection is to go for a set of views from very large search spaces that have the potential to minimize query processing costs and view maintenance. Formally, it can be stated as follows: Provided with a set of queries Q and a quantity S, the problem of the materialized view is to identify those set of views M, that tends to minimize the total maintenance cost and the total query response time while maintaining the special condition that the total size $(M) <$ size (S), that is, space occupied by M is less than S. This is considered as one of the most essential techniques in order to speedup the query answering in a warehouse application as data warehousing is a query-driven approach. The materialization of all views is not possible given its query-intensive nature. So, in order to assess the effectiveness of the proposed SFLA, comprehensive performance study has been conducted. The transaction processing performance

TABLE 8.1

Query Running Time Comparison

Query	GHA	GA	SFL
10	1.19 h	16.72 min	1.34 min
20	5.26 h	30.46 min	5.28 min
30	10.57 h	51.85 min	13.61 min
40	20.83 h	1.43 h	21.52 min

council (TPC)- D (Decision Support) (TPC-D) benchmark dataset has been used for the same (Table 8.1). The results can be summarized as follows:

1. The proposed SFLA performs better as compared to greedy heuristic algorithm (GHA) and the popular GA.
2. The SFLA has the capability to find the global optimal solution within very less amount of running time.
3. The SFLA can also avoid being falling in the local optima. It further merges the benefits of mimetic algorithm and PSO algorithm to improve the quality of solutions.

In [10], a discrete shuffled frog leaping algorithm (DSFLA) is proposed to solve the RNA secondary prediction problem. In any biological system RNA is considered to one of the crucial elements. It is hard to get RNA 3D-structutres using experiment methods such as X-ray crystallography and NMR because a very large amount of structural information is available. In addition, techniques mentioned are time taking and expensive too. Various researches have concluded that obtaining RNA tertiary structure directly from its primary counterpart is a complicated task. So, predicting the tertiary structure of RNA while determining the secondary structure of RNA is often a step forward. In order to explore the RNA secondary structure in combinational space of stems, a new individual location updation rule and search space must be defined as per the uniqueness of RNA folding. For this, new SFLA approach called DSFLA is used for finding RNA structures with minimum free energy.

8.2.1 Discrete Shuffled Flog Leaping Algorithm

1. Initialize each solution X_i of the swarm with evaluation
2. Calculate fitness of the frogs, rank them, and partition them into memeplexes, update X_g
3. Perform mimetic evolution within each memeplex
   a. *Step 1*: X_w (worst frog within the memeplex) produces and evaluates the newly created solution
   b. *Step 2*: Evaluate the new solution, if not better, replace X_b with X_g and go to Step 1
   c. *Step 3*: Evaluate the solution, if it is still not better, then replace X_w with a randomly compatible stems set
4. Output the best solution if the stopping criterion satisfied; else go to Step 2

[11] used SFLA to unravel the biclustering of microarray data, and mining coherent patterns from microarray data by proposing a multiobjective shuffled frog leaping biclustering (MOSFLB) algorithm. Many clustering methods have been rigorously used for the analysis of gene expression. Genetic-based grouping or clustering has the potential to spot set of genes having similar profiles. Though, clustering techniques make an assumption that related genes have similar expression patterns in all situations. As the dataset may contain many heterogeneous situations, the assumption is not good enough to proceed. Many clustering algorithms are known to work in the full dimension space. This tries to cluster the common points together by considering the value of each point in all the dimensions. However, the fact remains that relevant genes may not be essentially related in all conditions; as a result, the clustering may not be very efficient. As a result, a new proposed approach known as biclustering approach is used for grouping the genes simultaneously such that the gene subsets exhibit similar expression patterns. [11] proposes a novel multiobjective SFL biclustering framework in order to mine the biclusters from microarray datasets. The authors have focused on finding maximum biclusters such that they have lower mean squared residue and higher row variance. Consequently, a SFLA method is used to balance and manage the search process, whereas the sigma method is employed to find a better local guide in an objective space.

In [12], a quantum SFLA is presented. This approach trains the neural network using the quantum theory. The motivating factor is that the time-various functions are used as input and output of neural networks. In addition, aggregation operation has the potential to reflect the space aggregation function of the time-varying input signals. However, the computations are quite complex. So, in order to simplify the scheme, a quantum shuffled frog leaping algorithm (QSFLA) has been proposed. This algorithm uses Bloch spherical coordinates of qubits to express the individuals. The quantum individuals get changed using the quantum rotation gates, and the mutation of individuals is gained with Hadamard gates. A straightforward method is also proposed to determine the size, direction, and angle of quantum rotation.

8.2.2 Quantum Shuffled Frog Leaping Algorithm

1. Initialization and calculation of the Bloch coordinates
2. Transform the solution space and evaluate the fitness
3. Arrange fitness (descending order), and revise the global optimal solution
4. Update the worst individual in each memeplex
5. Perform mutation operation on the individual solutions
6. Output the result if the stopping criterion is satisfied; else go to Step 2

[13] authors proposed a combination of SFLA with an external optimization algorithm known to be a new hybrid optimization algorithm (SFLA-EO). This hybrid algorithm defines the application of an external optimization (EO) to SFLA. EO recently emerged as a new approach for a general-purpose local search optimization. The algorithm was proposed by Boettcher and Percus. EO has been applied fruitfully to various applications by removing the worst components in the suboptimal solutions. It has been applied in problems such as TSP, graph coloring, spin glasses, and so on. In this hybrid, shuffled

frog leaping algorithm-external optimization (SFLA-EO) algorithm, global search ability of SFLA ensures that the search converges quicker; on the other hand it utilizes EO ability of strong local search to get global optima by jumping out of local optima. On comparing the results of experiments with the standard SFLA, PSO, and EO over six well-known benchmark functions [25], including Ackley, Rastrigin showed that the proposed algorithm outperformed in terms of the speed of convergence, robustness, and stability. The algorithm also proved its effectiveness and superiority in solving continuous global optimization problems.

[14] proposed a modification on standard shuffled frog leaping algorithm (MSFLA). The algorithm adopts a new searching strategy. In this search strategy, simultaneous adjustment of frog's position is done in terms of the best individual present within the memeplex and the global best of population. In addition, for maintaining a balance between the global and local search ability an effect factor is introduced. In the original algorithm, the local best solution causes the algorithm to perform well in the early iterations, but the algorithm is shown to slow down or even become stagnant with increasing iterations. So, the proposed MSFLA adopts a new searching strategy with an objective to avoid local optimal solution. In this strategy, the position of the frog is adjusted using the knowledge of the best individual within the memeplexes and the global best of population simultaneously, and the last position of the frog is changed using the effect factor. In addition to this, the ergodicity as well as the diversity of the population can be enhanced by using the chaotic variables. Five benchmark functions each with 10, 20, and 30 variables are solved to accommodate different sizes of the problem. The better convergence speed is established with the simulation using MSFLA.

In an attempt to advance on the convergence speed, two-phase SFLA [15] is proposed. Various problems of optimization emerging from the domains of science, engineering, and management can be solved using stochastic search techniques. Due to this, these have been in focus of the subject for many researchers, scientists, and academicians to solve larger level of optimization problems. Due to their ease of implementation and simplicity they are peculiar over traditional techniques. These techniques require only supplementary understanding about the problem to solve the problems in an efficient way. The algorithm can be put as

1. The mixture of randomly generated method and opposition based learning (OBL) is used for the generation of an initial population. This is in contrast to using full initial population as random. To improve on the exploration properties of the algorithm and further for testing a large search space, the reasoning of the OBL is used. This phase is incorporated mainly to ensure an increased global convergence by uniformly distributing the population of solutions in the feasible region.

2. The searching mechanism for the worst frog is accelerated by using a scalar factor component. This component essentially improves the worst frog's position.

In [16], an improved SFLA based on a path planning technique is proposed. In this approach, in order to avoid local optimal solution problem that exist in the original algorithm, a new updating mechanism is engaged. This update mechanism is based on the median energy parameter. In addition, the fitness function of the algorithm is also modified to smooth the path generated. The globally best frog obtained on an itera-tion to iteration basis dictates the movement of the robot. The algorithm is represented below:

1. Initialization. Parameters, including the number of frogs F, the number of meme-plexes m, the maximum step size D max, and so on, are initialized randomly.
2. The initial population is generated. The current position of robotic is used to mark the center of the circle having an area A. The radius of this circle is the detecting radius of the robotic onboard sensors. Then the generated frogs are confined in this circle of area A.
3. Fitness of all the frogs is computed and then they are arranged in a descending order based on their fitness values.
4. The population is partitioned so that the frog occupies the m memeplexes.
5. Local searching is then performed. In this stage, the median strategy is used for the updation of the worst frog in every memeplex. This update is done in L iterations.
6. Shuffling process is performed. Once the local search is complete, frogs in the population shuffle are shuffled. The position of the optimal frog describes the next position of the robot.
7. Figure out if the target is achieved by the robot, if yes, then end of the path plan-ning task. Else, return to step 2.

In an attempt to verify the achievability and efficacy of the approach, simulations are done. The simulations are done in a static as well as a dynamic environment. The experimental results show that the proposed improved shuffled frog leap algorithm (ISFLA)-based method is quite effective and feasible irrespective of the environment considered.

In [17], an algorithm is proposed to solve the price-based unit commitment problem (PBUCP) under a deregulated environment using SFLA. Price/Profit-based unit com-mitment (PBUC) formulation considers the softer demand constraints and assigns fixed and transitional costs to the scheduled hours. The main objective of this approach is to make the best use of generating companies by scheduling the optimal allocation of gen-erating units. This problem is based on the forecasted price and the power demand. The motivation in solving PBUC problem is the maximization of the total profit of generation companies (GENCO) across the scheduling horizon. In an attempt to increase their own profit, GENCOs simulate the profit-based unit commitment by taking into consideration the forecasted demand and reserve prices in the markets. In the competitive power market, GENCOs run their own unit commitment schedule to maximize its profit. These forecasted data are of crucial importance to the GENCOs unit commitment problem since they are tak-ing the risk of committing their own units. The SFLA has been tested on 10 different unit systems over a scheduling period of 24 hour. Results were obtained for the optimum unit commitment schedule. Mega watt values were obtained for real power, hourly profit, and unit-wise profit of each generator unit in a given time frame. The total profit of the GENCO was also considered. It has been established from the observed results that the presented algorithm provides maximum profit and is computationally cheap (less time required by the algorithm). As such, such an approach can readily be used in real-time generation units.

With an ever generating documented data present on the web and the advancement of computational power, information retrieval (IR) has gained a promising importance in the act of retrieving documents from a large-scale document database. It has become a tentative research topic in recent years. The idea of extracting the discriminating features of the documents and their subsequent retrieval has become a critical issue. As the documented data are of very high dimensionality (dimensionality can be of the order of thousands), it is very much advantageous in transforming the high dimensionality documents onto a lower-dimensional subspace first. The fact remains that it is in a lower dimension space that the semantics of the document space becomes understandable. In addition, the classical query optimization algorithms available can then be applied on a document space in lower dimensional space.

In the same context, document query optimization algorithm is proposed in [18]; the same finds its compatibility in dealing with document query optimization problem. This novel algorithm is a combination of locality preserving projection (LPP) and SFLA. The LPP method is used for pattern recognition and is based on linear dimensionality reduction. LPP is different from latent semantic indexing (LSI) in the working as LSI seeks to mine the most representative features, LPP on the other end targets to maintain the local manifold structure while mining the most discriminative features present. The LPP can also be used as a linear approximation to the Eigen functions of the Laplace Beltrami operator on the document manifold. Thus, it has the potential to discover the nonlinear manifold structure to some extent. Reuters-21578 dataset (that consists of well-documented structures) is used in the simulations. Precision is used to evaluate the document retrieval performance. Precision is expressed as the ratio of the number of relevant documents retrieved in the top K (we empirically set $K = 10$) retrieved documents. For measuring the performance of the algorithm authors have also considered the precision metric. In the experiments, the performance of LPP–SFL, relevant feedback (RF) approach, and GA-based query optimization algorithms is compared.

In [19], authors presented a modification in SFLA specifically for solving the TSP by proposing a model for feasible travel route as a frog. TSP is a typical example of a combinatorial optimization problem. Formally, it is explained as: finding a minimum cost path of visiting all the cities once provided with the number of cities and the cost of reaching it. To get better on the class of initial population, simple near neighborhood rule is used to generate some solutions (frogs). For updating the worst frog position, the revised order crossover inversion mutation (ROXIM) operator is developed. The operator uses the key features of the two, that is, the revised order crossover (ROX) and the inversion mutation (IM) operator to optimize the local search operation. The experiments are carried out on Br17, P43, and Berlin52 for demonstrating the performance of the modified SFLA and the ROXIM operator. Experimental results confirm the following facts:

1. The TSP problem can be handled using the modified SFLA.
2. Algorithm's robustness can be improved by the ROXIM operator's contribution.

[20] demonstrated fast SFLA (FSFLA) to overcome the deficiencies such as slow speed, low optimization precision, and lack of trapping in local optimum that exists in the original SFLA algorithm. In the initial phase, each individual of subgroups is made to learn from the group extremum and the subgroup extremum. The *hit-wall* method is used for controlling the boundaries.

Further enhancement in the speed of this algorithm can be achieved by sorting and grouping all the individuals at regular intervals. Subsequently, in order to retain most of the individuals and take maximum advantage of the individual's information, a small fraction of individuals are randomly generated.

SFLA works with the learning of only worst individuals. The best individual is used for the learning of worst individual in any memeplex. If no improvement is achieved then the global best individual is used for the learning of the worst one. In FSFLA speedup is achieved by providing simultaneous learning from the local best individual in subgroup and the global best solution to each frog rather than only worst frog as in SFLA.

The traditional SFLA incorporates the replacement of worst frog with the random frog in case of not gaining any improvement after learning from the local and the global best individual. Although, while adding the diversity to the population using the traditional approach, there is a flaw as well, that is, it completely boycotts the beneficial information that was present in the worst individual. The FSFLA on the other hand updates the ith frog when it does not improve after the update.

A quantum binary shuffled frog leaping algorithm (QBSFLA) [21] is proposed. This strategy integrates the concept of SFLA and the quantum evolutionary algorithm. The overview is as follows:

First, the superposition state characteristic of quantum allows separate individual to express more states, and the probability expression makes individual's states to be expressed with certain probability. This probability term has got a potential to increase the population diversity.

In the next step, the phase of the quantum bit is regulated by SFLA. This is done to achieve a balance between the local and the global search.

The performance improvement of the algorithm is established using the experimental results of three 0–1 knapsack problems. The improvement is in the terms of increased convergence speed, global search ability, and stability.

A SFLA-based algorithm is proposed to solve the flexible job shop scheduling problem (FJSP) [22]. Distinct from the traditional approach, in FJSP not only operations can be processed in any machine but also they can be processed in different orders. This essentially means that the FJSP can be transformed into two subproblems, that is, problem of the machine assignment and the operation scheduling. The objective of the first subproblem is the assignment of machine from machine pool to each operation of a job and the second subproblem of operation scheduling seeks to settle on the processing order of operation on each machine. Considering the above-mentioned aspects, a reasonable schedule can be obtained [2]. The prerequisite for solving the FJS is finding a reasonable (if not optimal) schedule by assigning a proper machine to each operation. Certain scheduling objectives are guaranteed with this and also lay more complexity of the FJSP problem than the classical JSP as it contains two subproblems.

Following approach has been used:

1. First, hybrid scheme is implemented in SFLA, in which the order sequence and machine assignment operations are demonstrated using the two sequences.

2. Second, a bilevel crossover scheme is integrated in the framework of the mimetic search. This scheme is used as a local search strategy. The searching ability of the modified SFLA is such that it strikes a good balance between the exploration and the exploitation.

3. In addition, the influence of the parameter setting is thoroughly examined and based on the observations, optimal values of parameters are assigned.

The optimality of SFLA algorithm for solving FJSP is verified with the computational results and comparisons.

In order to do away with the shortcomings such as slow speed, low optimization precision, and lack of trapping in local optimum that exists in original SFLA algorithm, a CSFLA [23–25] is proposed. This approach has gained insight because of the artificial fish-swarm algorithm (AFSA). In AFSA optimization speed is accelerated by the exploitation of the follow behavior of fish-swarm, and the swarm behavior is used to develop the capacity of moving out of the local extremum.

This work embodies AFSA into SFLA such that the follow property is there to boost the convergence velocity of population (in local depth search stage), and swarm property comes handy to leap out the local extremum, and improves on the convergence of population. Gaussian mutation is used to implement the feeding mechanism. The simulation results show that the modified algorithm yields better performance in preventing premature convergence and effective global search capability.

Pseudo Code for SFLA Algorithm

Begin
Generate random population of P solutions (frogs)
Calculate fitness(i), for each individual$_i$ in the population
Sort the individuals in the descending order of their fitness values
Then determine the global best solution X_{gb}
Divide Population P into m memeplexes
For each memeplex, do:
  For each mimetic iteration, do:
    Determine the local best (X_{lb}) solution and the worst frog (X_w) present
    Change in the frog position is given by $D_i = rand() \ x \ (X_{lb}-X_w)$
    If the position of the frog improves then, do:
    Update new position as $X_w = X_w + D_i$
    Else
    Change in the frog position $D_i = rand() \ x \ (X_{gb} - X_w)$
    If the position of the frog improves, then do:
    Set new position $X_w =$ as $X_w + D_i$
    Else
    New position $X_w = rand()^*D_{max} + rand()^*D_{min}$
  End
End;
Combine the evolved memeplexes obtained
Arrange the individuals in P in descending order of their fitness
Confirm if termination condition is satisfied;
End;
End;

Figure 8.1 shows the flowchart of SFLA algorithm.

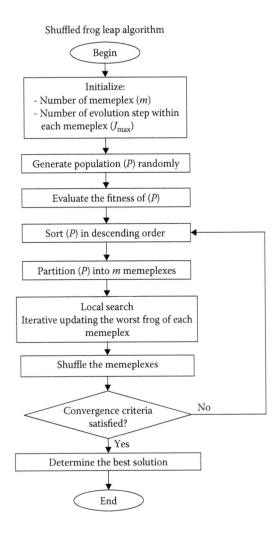

FIGURE 8.1
Flowchart of the SFLA.

Solved Questions

1. Compare the performance of SFLA when used for solving TSP and QAP.

Solution

For comparing the performance of the algorithm for solving the two problems, the algorithm has been implemented in C++ on Code Block IDE.

For the TSP problem, Berlin52 dataset is used. The results obtained are tabulated in Table 8.2.

In order to solve the QAP problem, three different datasets have been used. The results are tabulated in Table 8.3.

TABLE 8.2

Simulation Results for TSP Problem Using BERLIN52 Dataset

Given Optimal	Mean over 5 Runs	Worst Obtained	Best Obtained	Error
7,500	9,627	10,453	8,979	2,127

TABLE 8.3

Simulation Results for QAP Problem Using Respective Datasets

Dataset	Number of Iteration	Given Best	Mean over 5 Runs	Best Obtained	Worst Obtained	Error
Nug12	300	578	593	582	610	14
Esc16i	40	14	14	14	14	0
Kra32	100	8,8700	9,4076	9,2476	9,6460	5,376

2. If the local best solution and the worst frog present are given as values 17 and 7, respectively, then determine the change in the frog position.

Solution

For calculation of change in the frog position, we find a random variable in the range [0,1]. In our case, the number was 0.17.

Using

$$D_i = rand() * (X_{lb} - X_w)$$

we have

$$D = 0.17 * (17 - 7) = 1.7$$

Unsolved Questions

1. Compare the complexity of TSP with that of QAP.
2. Discuss the concept of memeplex in context to SFLA.
3. Can the number of evolutions within a memeplex affect the convergence of the algorithm? Explain.
4. What factors prevent the SFLA algorithm from converging adequately? Discuss the ways to mitigate their effect.
5. What elements differentiate the BSFL from IBSFL? Why is the latter *improved*?
6. Discuss the rationale behind using the bilevel crossover in SFLA.

7. What steps need to be adopted to strike a perfect balance between the exploration and exploitation in SFLA?

8. Comment on the inherent limitations of SFLA algorithm.

9. Write a short note on ROXIM operator.

10. Comment on the usefulness of SFLA algorithm in the field of artificial intelligence.

References

1. R. Balamurugan, Application of shuffled frog leaping algorithm for economic dispatch with multiple fuel options, *International Conference on Emerging Trends in Electrical Engineering and Energy Management*, IEEE, Chennai, India, 2012.

2. M. Barati and M. M. Farsangi, Solving unit commitment problem by a binary shuffled frog leaping algorithm, *IET Generation-Transmission & Distribution*, November 2013.

3. M. Gomez-Gonzalez and F. Jurado, Personalized E-learning using shuffled frog-leaping algorithm, *Global Engineering Education Conference (EDUCON)*, IEEE, Marrakech, Morocco, 2012.

4. I. G. Sardou, M. Banejad, R. Hooshm and A. Dastfan, Modified shuffled frog leaping algorithm for optimal switch placement in distribution automation system using a multi-objective fuzzy approach, *IET Generation-Transmission & Distribution*, 2011, 6(6), 493–502.

5. I. Hassanzadeh, K. Madani and M. A. Badamchizadeh, Mobile robot path planning based on shuffled frog leaping optimization algorithm, *IEEE Conference on Automation Science and Engineering Marriott Eaton Centre Hotel*, Toronto, ON, August, 2010, vol. 6, pp. 21–24.

6. T. H. Huynh, D. H. Nguyen, Fuzzy controller design using a new shuffled frog leaping algorithm. In *Industrial Technology, 2009, IEEE International Conference on 2009* February 10, pp. 1–6. IEEE.

7. S. Jaballah, K. Rouis, F. B. Abdallah and J. B. Tahar, An improved shuffled frog leaping algorithm with a fast search strategy for optimization problems. In *Intelligent Computer Communication and Processing (ICCP), 2014 IEEE international conference on 2014* September 4, pp. 23–27. IEEE

8. Y. Li, J. Zhou, J. Yang, L. Liu, H. Qin and L. Yang, The chaos-based shuffled frog leaping algorithm and its application, *International Conference on Natural Computation*, IEEE Computer Society, Washington, DC, 2008.

9. X. Li, X. Qian, J. Jiang and Z. Wang, Shuffled frog leaping algorithm for materialized views selection, *International Workshop on Education Technology and Computer Science*, Wuhan, China, 2010, vol. 2, pp. 7–10.

10. J. Lin, Y. Zhong and J. Zhang, Discrete shuffled flog leaping algorithm for RNA secondary structure prediction, *International Conference on Natural Computation*, IEEE, Shanghai, China, 2011, vol. 7, pp. 1489–1493.

11. J. Liu and Z. Li, Multi objective optimization shuffled frog-leaping biclustering, *International Conference on Bioinformatics and Biomedicine Workshops*, IEEE, 2011, pp. 151–156.

12. L. Liu and Q. Zhang, Training and application of process neural network based on quantum shuffled frog leaping algorithm, *International Conference on Computer Science and Network Technology*, IEEE, Dalian, China, 2013, vol. 3, pp. 829–833.

13. J. Luo, M. Rong and C. X. Li, A novel hybrid algorithm for global optimization based on EO and SFLA, *Industrial Electronics and Applications*, IEEE, 2009, pp. 1935–1939.

14. P. Luo, Q. Lu and C. Wu, Modified shuffled frog leaping algorithm based on new searching strategy, *International Conference on Natural Computation*, IEEE, Shanghai, China, 2011, vol. 7, pp. 1346–1350.

15. B. Naruka, T. K. Sharma, M. Pant, J. Rajpurohit and S. Sharma, *Two-phase Shuffled Frog-Leaping Algorithm*, IEEE, 2012.

16. J. Ni, X. Yin, J. Chen and X. Li, An improved shuffled frog leaping algorithm for robot path planning, *International Conference on Natural Computation*, IEEE, Xiamen, China, 2014, vol. 10, pp. 545–549.

17. S. Kumar, T. Yenkatesan and M. Y. S. Vullah, *Price Based Unit Commitment Problem Solution Using Shuffled Frog Leaping Algorithm*, IEEE, 2012.

18. Z. Wang, Q. Zhang and X. Sun, *Using LPP and SFL for Document Query Optimization*, IEEE, 2009.

19. M. Wang and W. Di, A modified shuffled frog leaping algorithm for the traveling salesman problem, *International Conference on Natural Computation*, IEEE, Yantai, China, 2010, vol. 6, pp. 3705–3708.

20. L. Wang and Y. Gong, A fast shuffled frog leaping algorithm, *International Conference on Natural Computation*, IEEE, Shenyang, China, 2013, vol. 9, pp. 369–373.

21. L. Wang and Y. Gong, Quantum binary shuffled frog leaping algorithm, *International Conference on Instrumentation, Measurement, Computer, Communication and Control*, IEEE, Shenyang, China, 2013, vol. 3, pp. 1655–1659.

22. Y. Xu, L. Wang and S. Wang, *An Effective Shuffled Frog-leaping Algorithm for the Flexible Job-shop Scheduling Problem*, IEEE, 2013.

23. X. Zhang, F. Hu, J. Tang, C. Zou and L. Zhao, A kind of composite shuffled frog leaping algorithm, *International Conference on Natural Computation*, IEEE, Yantai, China, vol. 6, pp. 2232–2235, 2010.

24. E. Afzalan, M. A. Taghikhani and M. Sedighizadeh, Optimal placement and sizing of DG in radial distribution networks using SFLA. *International Journal of Energy Engineering*, 2012, 2(3), 73–77.

25. Z. Zhen, Z. Wang, Z. Gu and Y. Liu, A novel memetic algorithm for global optimization based on PSO and SFLA, *Lecture Notes in Computer Science*, 2007, 4683, pp. 126–136. Springer-Verlag, Berlin, Germany.

9

Brain Storm Swarm Optimization Algorithm

9.1 Introduction

In this chapter, we will explore a different nature-inspired algorithm, inspired by the process of human brainstorming for solving any complicated problem. All of us are acquainted with the fact that, when a complex problem comes in front of us, and a single person is not able to solve, then a group of people with different qualities come together and do brainstorming. In this way, the probability of solving that problem is increased. In the process of brainstorming, great ideas come out and their collaboration with each other helps to generate better ideas for solving the problem. Brainstorm optimization algorithm has been applied for solving very challenging and hard problems by randomly generating the ideas or individuals. This is an optimization algorithm that is developed on the basis of human ideas generation process, and as humans are the most knowledgeable and intelligent creatures in the world, this algorithm is better than the other nature-inspired algorithms in various aspects.

9.2 Brain Storm Optimization

Osborn's four rules for generating ideas are as follows:

1. *Focus on quantity*: It is assumed that if the number of generated ideas is more, then there will be the higher chance of getting an efficient and radical solution. This rule aims to make problem-solving easy with the concept that quantity breeds quality.

2. *Withhold criticism*: Rather than criticizing the generated ideas, participants of brainstorming should add more and more ideas and should criticize the ideas at the later stage, by which participants participating in brainstorming will produce unusual and creative ideas, with a higher probability.

3. *Welcome unorthodox ideas*: By viewing problems from different perspectives and suspending assumptions, participants can generate unusual ideas. These solutions are welcomed for getting more and more ideas.

4. *Cross-fertilize*: According to this rule, better ideas may be obtained by combing good ideas, as recommended by the famous saying "1 + 1 > 2."

Steps in the process of brainstorming are as follows:

Step 1: Group of people from different background come together for brainstorming process.

Step 2: Produce as many ideas as possible according to the rules of Osborn.

Step 3: Let us say 2 or 4 clients act as the owners. Keep a number of ideas from each of the owner as better ideas.

Step 4: Ideas chosen in Step 3 are given higher probability than any other generated ideas. Now produce more and more ideas or thoughts according to the rules of Osborn.

Step 5: Keep the owners to select a number of better ideas or thoughts generated in Step 3.

Step 6: Arbitrarily choose an object and the appearance and functions of that object are used as clues.

Step 7: Keep the owners to choose a number of better ideas.

Step 8: An optimal solution for the problem can be obtained by merging the generated ideas.

9.2.1 Brain Storm Optimization Algorithm

1. *Initialization*: Arbitrarily produce n possible solutions or individuals
2. *Grouping*: Cluster these n solutions into m groups or clusters
3. Evaluate or calculate the fitness of n individuals
4. Now give ranking to solutions in each group or cluster and the best individual or solution in each group or cluster is recorded as the center of that particular cluster
5. *Replacing*: Arbitrarily produce a value $\in [0,1]$
   a. If generated value is less than a fixed value p_replace
      i. Arbitrarily choose a cluster or group center
      ii. Arbitrarily generate a solution and replace the chosen cluster center with the generated solution
6. *Population update*: Produce new solutions
   a. Arbitrarily produce a value $\in [0,1]$
   b. If generated value is less than a fixed value p_one
      i. Arbitrarily choose a cluster k with probability p_k
      ii. Arbitrarily produce a value $\in [0,1]$
      iii. If generated value is less than a fixed value p_one_center
   Select group center and add a random value to this group center to create a new idea.
      iv. Or else, arbitrarily choose an idea from this group and add random value to the idea to create a new idea
   c. Or else, arbitrarily choose two groups to create a new idea
      i. Produce an arbitrary value
      ii. If generated value is less than a fixed value p_two_center, the two group centers are joined and then an arbitrary value is added to create a new idea

  iii. Or else, two ideas from both the chosen groups are arbitrarily elected to be combined and then an arbitrary value is added to create a new idea

 d. The new produced idea is compared with the existing idea with the similar idea index; better idea is kept as a new idea

7. If n new ideas have been produced, go to Step 8; else, go to Step 6

8. Finish the procedure if fixed numbers of iterations are done; else, go to Step 2

The working of the brain storm swarm optimization (BSO) is also represented by the flow-chart given below (Figure 9.1) [1]:

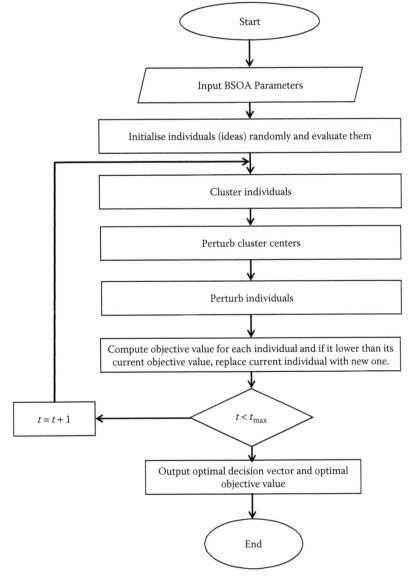

FIGURE 9.1
General flow of BSO.

9.3 Related Work in Brain Storm Optimization and Other Contemporary Algorithms

BSO algorithm is a relatively new and widely used swarm intelligence [2] algorithm that was first proposed by Shi in 2011 [3,4]. However, unlike other swarm intelligence algorithms such as the ant colony optimization (ACO) [5,6], honey bee optimization (HBO), bacterial forging optimization (BFO), and particle swarm optimization (PSO) [7,8] that imitate the joint behaviors of animals such as ants, bees, birds, and bacteria, BSO algorithm is stimulated by the collective behavior of more intellectual organisms, that is, human beings. It is natural to be expecting that BSO algorithm should be better than other swarm intelligence algorithms, as it emulates the smartest species in the globe.

As discussed earlier, whenever humans face a complicated problem, which a single person may find difficult to solve, a group of people collectively do brainstorming and then the problem is easily solved. These people, frequently with diverse qualities participate in the brainstorming process that help these people to interactively work together to produce great ideas. Using this approach, most of the problems can be solved. *Shi* has designed an efficient BSO algorithm by imitating this brainstorming process found in human beings for solving difficult problems [3]. Zhan et al. [9] proposed a new BSO algorithm, named modified BSO (MBSO) by customizing the grouping operator and the creating operator, to improve the algorithm efficiency. Zhou et al. [10] has developed a different versions of BSO by modifying the creating operator. Xue et al. [11] has introduced the Cauchy random noise in the creating operator.

The robot motion planning (RMP) issue is to uncover a *without impact way* from the beginning position to the target position for robots that evades obstructions, has been inquired about beginning from the mid of 1970s. Numerous fantastic methodologies have been proposed for tackling this issue [12]. These current excellent routines are the varieties of few general methodologies: cell decomposition, RoadMap, mathematical programming, and potential fields. These methodologies are not fundamentally unrelated, so consolidations of these methodologies are frequently utilized for creating movement planners. RMP issue is an NP-Hard issue, so for minimizing complexities and high computational expenses of fantastic systems, heuristic strategies have been produced over the past two decades, essentially for high degrees of flexibility.

In heuristic calculations, it is not ensured to discover an answer, yet in the event that an answer is discovered; it will be carried out much quicker than deterministic strategies. Principle meta-heuristic systems proposed in RMP are genetic algorithms (GAs), tabu search (TS), simulated annealing (SA), ant colony optimization (ACO), and particle swarm optimization (PSO). Survey of the provisions of exemplary techniques and these abovementioned heuristic methodologies in RMP is exhibited in [13,14].

One approach for robot path planning is proposed in [15] using PSO with mutation operator. An obstacle avoidance algorithm for path planning of soccer robots using PSO algorithm is developed in [16], smooth path planning for a mobile robot using stochastic PSO algorithm is developed in [17]. Probabilistic algorithms are the robot motion planning algorithms; other successful heuristics include rapidly exploring random trees (RRT) and probabilistic roadmap method (PRM) [18].

As discussed in Chapter 2, PSO algorithm [19], was first proposed in 1995 in a population-dependent technique motivated by flocks of birds and fish schools. ACO [20], proposed in 1996, is dependent on the patterns of ants in their search for meal. ACO has been discussed in detail in Chapter 4.

Simulated annealing [21] was initially identified in 1983, and is analogical to annealing in metallurgy. Whenever a material is heated, its atoms get freed from their initial position in the structure where they were restricted in, and have the ability to randomly adjust their positions trying to search out for arrangement with reduced internal energy as compared to that of their initial state. The capability of atoms to transform their position reduces with the temperature until almost all of the atoms stay in their predetermined positions. The algorithm operates the same. In every single iteration, a different position is found by changing the existing position, in case the new one is a better choice, it is approved. When a new position is even worse as compared to the existing one, it is still approved with the probability determined by the system's temperature. The temperature reduces with time, and so worse positions are accepted with good probability at the start of the search, enabling exploration of big section of the subspace. Later, the search is targeted on a smaller section of the search space. Simulated annealing is helpful in case of structural optimization or optimization of water distribution or transportation systems. The GA [22] was developed in 1975 by J. H. Holland. The concept of GA is discussed in detail in Chapter 3.

Differential evolution (DE) [23], proposed in 1997 is a population-centered evolutionary algorithm that is useful for the optimization of real functions or real parameters. The population of candidate solution is employed for the optimization, many new candidate solutions are made by composition of other candidate solutions from the population based on a simple and easy method and the perfect or the ideal solutions are kept in population, whereas all others are eliminated. DE showed to execute the job much better than the GA or simulated annealing for many problems. Additional benefits of DE are effortless and simple implementation and very less parameter tuning is needed.

Artificial immune systems (AIS) [24] were initially proposed as the models of the immune system. Since that time, studies have also targeted on resolving computation issues; and even today, AIS are carefully associated with immunology, due to the fact that the more superior and the better knowledge of biological immune systems, much more is the opportunity to employ the knowledge into developing a more effective and powerful artificial system.

Decentralized cellular evolutionary algorithms [25] are similar to evolutionary algorithms or GA. In GA, each individual in the population can get to know and communicate with each other, whereas in decentralized cellular evolutionary algorithms there will always be a described framework in the population. Individuals form groups or cells, and also the individuals who are within the same cell can communicate. This type of structure assists in preserving the diversity.

Robot motion planning or path planning in known/static and dynamic environments is possible for autonomous mobile robots [26]. In [27], it is assumed that the target is moving, and the workspace includes both types of obstacles: (1) static and (2) moving. A method has been proposed using the PSO algorithm for motion planning of autonomous mobile robots. Simulation results show that the total path is minimized and local optimums are avoided.

Multiagent routes planning [28] might be thought of as the essential building piece for utilizing a viable multiagent model skilled for collaborating with the actuality. A global path planner is good in generating a fully optimized path, but weak in reacting to not a known hindrance. In contrast, a local/reactive navigation technique is effective in dynamic and also initially undiscovered workplace, but specifically inefficient in complicated environment. Hence, a hybrid navigation method could be used as in [29].

9.4 Hybridization of Brain Storm Optimization with Probabilistic Roadmap Method Algorithm

In this section, we present the hybrid model consisting of the BSO algorithm and the PRM. BSO method is used as a global planner and PRM algorithm as the local planner, because of success of the PRM algorithm in high-dimensional spaces. The combination of these two algorithms is very efficient. Both the algorithms are compliment of each other and speedup the path planning process. An environment is considered, in which obstacles are static (static environment) and the source and target are in the form of 2D coordinates. An environment is said to be known if the positions of obstacles are known in advance and initial and target positions are also given. BSO algorithm has been applied for solving very challenging and hard problems by randomly generating the ideas or individuals.

The Following assumptions are made for the Algorithm:

- Initial and target location of a robot in the form of 2D points is known in advance.
- It is assumed that the obstacles are circular, rectangular, and triangular in shape.
- All obstacles are static, that is, their positions are given in advance.

Robot has some sensing range, so this algorithm generates initial points with respect to the robot's start location and robot's sensing range. An algorithmic diagram for avoiding static obstacles is shown in Figure 9.2. In this figure, initialization, grouping, replacing, and population updates are same as in general BSO algorithm. For getting a path from the source to the destination that avoids obstacles, a modification step is added as shown in Figure 9.2. In each iteration of this algorithm, the following steps are performed.

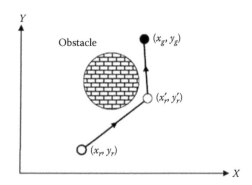

FIGURE 9.2
Obstacle avoidance using BSO.

1. *Initialization*: In the initialization step, BSO algorithm arbitrarily generates N ideas ($Xi = [x_{i1}, x_{i2} \dots x_{iD}]$, where $1 \leq i \leq N$, where N is the size of the population and D is the dimension of the problem) in the exploration space as the population. Here, in BSO algorithm we consider individuals as ideas like we have named individuals as particles in PSO algorithm and called individuals as ants in ACO algorithm.

2. *Grouping operator*: In the grouping step, BSO algorithm uses K-means clustering method for grouping individuals in clusters. N ideas are grouped into M different clusters. The best idea in each cluster is termed as cluster center.

3. *Replacing operator*: In the replacing step of BSO algorithm, p_replace parameter controls the operation. The value of p_replace is between 0 and 1. An arbitrary value $\in [0,1]$ is generated. If this randomly generated value is smaller than p_replace parameter, then replacing the operator randomly selects a cluster and replaces the cluster center with a randomly generated idea/individual.

4. *Population update*: This step is also named as creating operator step. In this step, BSO algorithm creates N new ideas that are generated on the basis of current ideas. For creating a new idea $Y_i = [y_{i1}, y_{i2} \dots y_{iD}]$ ($1 \leq i \leq N$), first, the BSO algorithm checks whether the newly generated idea Y_i is based on one cluster or based on two clusters. P_one parameter is used for controlling this. The value of this parameter is in between 0 and 1. If an arbitrarily generated value is less than p_one, then Y_i value is created using one cluster. Else, Y_i value is created using two clusters. For the first condition, cluster j is to be selected by a Roulette strategy. If there are more ideas in a cluster, the chances that it will be selected will be more. Probability of selecting cluster j can be written as

$$P_j = |M_j| / N \quad |M_j| \text{ are ideas in cluster } j \tag{9.1}$$

And for second condition, instead of selecting the two clusters according to Equation 9.1, BSO algorithm arbitrarily picks two clusters $j1$ and $j2$ from M clusters. After selecting cluster(s), BSO algorithm uses p_one_center $\in [0,1]$ (if based on one cluster) parameter or a parameter p_two_center $\in [0,1]$ (if based on two clusters) so that BSO can conclude, whether the created new idea Y_i is based on random idea(s) or cluster center(s) of the cluster(s). If the selected idea is X, then we can write the new idea Y_i as

$$y_{id} = x_d + \xi_d * N(\mu, \sigma)_d \tag{9.2}$$

Here, d is the dimension, $N(\mu, \sigma)$ is the Gaussian random value that has variance σ, μ is the mean, and ξ is a coefficient, which is used for weighting the contribution of Gaussian random values that we can calculate from the below formula:

$$\xi = logsig\left(\frac{(0.5 \times g - G)}{k}\right) * random(0, 1) \tag{9.3}$$

Where logsig() is known as logarithmic sigmoid transfer function $\in [0,1]$, g is known as current generation number, and k is a parameter used for changing the slope of logsig() function, G is the maximum number of generations. When BSO algorithm creates a new idea Y_i using two clusters, it first combines the two ideas from the clusters $j1$ and $j2$ and then uses Equation 9.2 to generate Y_i.

The combination equation can be written as

$$X = R * X1 + (1 - R) * X2 \tag{9.4}$$

Here R is a random value $\in [0,1]$, $X1$ and $X2$ are ideas from clusters $j1$ and $j2$. Using Equation 9.4, X is calculated and then Equation 9.2 is used and Y_i is calculated. After that, the BSO evaluates Y_i and replaces it with X_i if Y_i is more fit than X_i. In this way, a new idea-generating process is repeated for all the $1 \le i \le N$, and the complete population is updated.

5. *Modification step*: After population update step, modification step occurs. In the first iteration, best particle/2D point that is selected from the updated population is the part of final trajectory/path if this point does not collide with any of the obstacles. The best particle from the updated population is selected through the fitness function F from Figure 9.3.

$$F = \sqrt{\left(x_r - x_r'\right)^2 + \left(y_r - y_r'\right)^2} + \sqrt{\left(x_r' - x_g\right)^2 + \left(y_r' - y_g\right)^2} \tag{9.5}$$

The best particle from the updated population in iteration i is best_i. If the best_i collides with any of the obstacles, ignore it and select another best_i point from the set of particles. If new best_i does not collide with any of the obstacle, check whether the line joining best_i–1 and best_i does not collide with any of the obstacle. If the line does not collide with any of the obstacles then this will be the part of final trajectory. If this line collides with any of the obstacles, then apply probabilistic roadmap method between best_i–1 and best_i, get the path using the PRM algorithm between these points, and merge it with the previous path, see Figures 9.4 and 9.5.

BSO algorithm uses grouping operator, replacing operator, and creating operators to generate several possible ideas [30]. BSO mainly involves three control parameters in these operations: (1) *p_replace* is used to control the replacing operator, (2) *p_one* is used to control the creating operator to create new ideas between the clusters, and (3) *p_center* (*p_one_center* and *p_two_center*) is used to control the cluster center or random ideas to

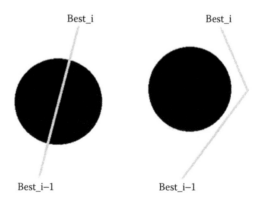

FIGURE 9.3
Path using a PRM algorithm.

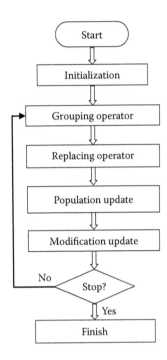

FIGURE 9.4
Flow diagram of a modified BSO.

1.	N: = {};
2.	E: = {};
	repeat
3.	q = random configuration from C;
4.	If q is free (i.e. no collision) then
	begin
5.	add q to N;
6.	Choose subset N_q of N with candidate neighbors for q;
7.	for all q′ in N_q sorted ascending by D(q′, q) do
	begin
8.	if local planner can connect q with q′
9.	then add (q,q′) to E
	end;
	end;
10.	Until limit is reached;

FIGURE 9.5
Pseudocode of PRM.

TABLE 9.1

Parameter Values in Standard BSO

Parameters	Values Taken from Reference 29
1. *p_replace*	0.2
2. *p_one*	0.8
3. *p_one_center*	0.4
4. *p_two_center*	0.4

```
Source Destination 0.5 0.5 9 9
Circle 0 0 0.5
Circle 5 2 1
Circle 1 1 0.3
Circle 1 2 0.3
Circle 2 2 0.3
Circle 4 4 1
Circle 4 7 1.5
Rectangle 1 3 2 2
```

FIGURE 9.6

Environment1 Input_File.Txt.

create new ideas. Replacing operator is controlled by p_replace parameter, which was proposed to be 0.2 in the study of Shi. p_one value was set to 0.8. The investigation results also show that *p_one_center* = 0.4, this value is very promising on most of the functions (Table 9.1).

The code is written in C++ and developed with the help of code block IDE. An output MATLAB® file is automatically generated using C++ code. In this MATLAB file, obstacles and the path avoiding these obstacles are created. Output is shown for different environments. Here, an input file is shown in Figure 9.6. In this input file, start and destination points are given. Positions of the obstacles are also given. First line shows the source and destination coordinates. First two values are x and y coordinates of start and last two values are x and y coordinates of target. This is the format of input file because this information has been extracted from the file and is used in the algorithm. Outputs for different environments are shown in Figures 9.7, 9.9 and 9.11 and their environment files are shown in Figures 9.6, 9.8, and 9.10. Here, black dots represent the particles generated in the BSO algorithm.

Format for triangle obstacle is	Triangle	X1 Y1 X2 Y2 X3 Y3
Format for circle obstacle is	Circle	X Y radius
Format for rectangle is	Rectangle	X Y width height

For comparing BSO algorithm's performance, PSO, DE, and PRM methods are used. In PRM method, first it constructs the probabilistic roadmap, and then it is searched by the

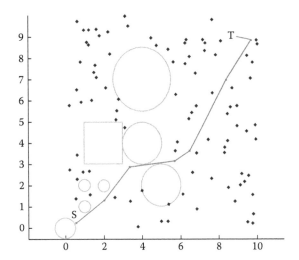

FIGURE 9.7
Path generated by algorithm in E_1.

Source Destination 0.5 0.5 9 9
Rectangle 1 3 2 2
Circle 0 0 0.5
Circle 5 2 1
Circle 1 1 0.3
Circle 1 2 0.3
Circle 2 2 0.3
Circle 4 4 1
Circle 4 7 1.5
Circle 10 10 .5
Triangle 8 3 12 3 6 6

FIGURE 9.8
Environment2 Input_File.Txt.

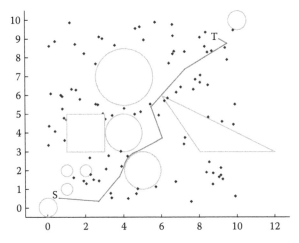

FIGURE 9.9
Path generated by algorithm in E_2.

Source Destination 0.5 0.5 9 9
Circle 1 2.5 1.2
Rectangle 1 2.5 2 2
Rectangle 4 2 2 2
Rectangle 7 3 2 2
Rectangle 5 5 2 2
Circle 7 8 1.2
Rectangle 6 7 2 2
Triangle 3 7 5 7 4 9
Triangle 8 6 10 6 9 8

FIGURE 9.10
Environment3 Input_File.Txt.

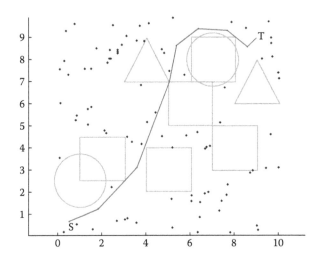

FIGURE 9.11
Path generated by the algorithm in E_3.

A* algorithm to get a start-to goal path. These methods are coded in C++ in code block IDE and programming quality style is same for all methods.

Figure 9.12 shows a graph of number of obstacles in different environments created with different numbers of obstacles. The running time comparison for all methods in different environments is shown in Figure 9.13. Results of the experiment are shown in Table 9.2. The standard deviation and the mean of runtime for all methods are calculated. Results show that the hybrid algorithm (BSO+PRM) is faster, efficient, and reliable than other algorithms.

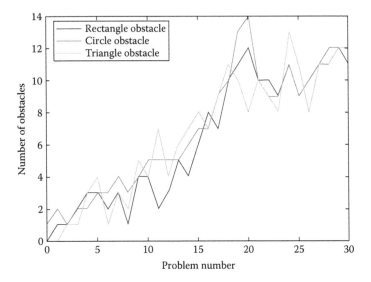

FIGURE 9.12
Number of obstacles in different problems.

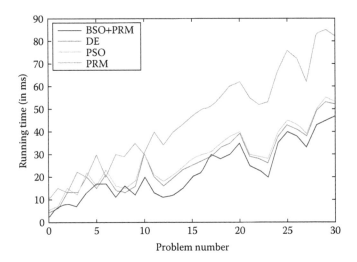

FIGURE 9.13
Running time comparisons.

TABLE 9.2

Means and Standard Deviations of Running Time

Algorithm	Mean Time	Standard Deviation
BSO+PRM	22.419	12.382
DE	26.741	12.981
PSO	28.322	13.115
PRM	41.903	20.507

9.5 Conclusion

In this chapter, the BSO algorithm is discussed in detail and its application in the field of robot motion planning is given. As we know that, this algorithm is inspired by the brain storming process found in human beings, and human beings are the most intelligent species on our planet, so this algorithm is one of the best nature-inspired algorithms as evident from the results mentioned in this chapter.

9.6 Future Scope

This algorithm is a human-inspired algorithm, and can be used for various real-life optimization problems, spanning across several disciplines, exploration, and task allocation problems, and can also be applied in multirobot system, where n robots and n targets are given in the workspace. Each robot has its own goal. These robots have to move to their targets without collision with obstacles (static or dynamic) and other robots in the environment.

Solved Examples

1. If the number of ideas created before the population update phase is 1000 and 233 ideas come from a cluster *Manager*, then compute the probability of *Manager*'s ideas being selected.

Solution 1

Given,

Number of ideas created before the population update phase (N) = 1000

Number of ideas coming from cluster *Manager* (M_{Manager}) = 233

Using,

$$P_{\text{Manager}} = \frac{\left|M_{\text{Manager}}\right|}{N}$$

$$P_{\text{Manager}} = \frac{233}{1000} = 0.233$$

Therefore, the probability that the idea generated from the cluster Manager is selected in the population update phase is 0.233.

2. Justify the logic behind *withholding criticism* in Osborne's rules.

Solution 2

The simple logic behind *withholding criticism* in Osborne's rules is that, if all the ideas, no matter how uncommon they may be, are welcomed during the initial phase, then the participants will produce more unusual and creative ideas, without any fear of criticism, which will help in exploring different possibilities for solving the problem, which could otherwise have not been introduced during brainstorming due to the fear of criticism.

Exercises

1. How is BSO different from other nature-inspired algorithms?
2. Contrast the notion of the shortest path with the optimally best path.
3. Distinguish between the requirements and constraints of global path planning with that of local path planning.
4. How is the population updated in BSO?
5. Comment on the suitability of Osborn rules in BSO.
6. How are *owners* manifested in a BSO algorithm? Comment on their usefulness.
7. Discuss the role of cross-fertilization phase in context of BSO algorithm.
8. Write down the various applications of the BSO algorithm.
9. Discuss the role of replacing operator in BSO.
10. Describe the motivation behind the use of BSO.

References

1. A. R. Jordehi, Brainstorm optimisation algorithm (BSOA): An efficient algorithm for finding optimal location and setting of FACTS devices in electric power systems. *International Journal of Electrical Power & Energy Systems*, 2015, 69, 48–57.
2. J. Kennedy, R. C. Eberhart, and Y. H. Shi, *Swarm Intelligence*, Morgan Kaufmann, San Mateo, CA, 2001.
3. Y. Shi, Brain storm optimization algorithm, *Proceedings of the 2nd International Conference on Swarm Intelligence*, Springer, Berlin, Germany, 2011, pp. 303–309.
4. Y. Shi, An optimization algorithm based on brainstorming process, *International Journal of Swarm Intelligence Research*, 2011, 2(4), 35–62.
5. W. N. Chen and J. Zhang, Ant colony optimization approach to grid workflow scheduling problem with various QoS requirements, *IEEE Transactions on Systems, Man, and Cybernetics C*, 2009, 39(1), 29–43.

6. Z. H. Zhan, J. Zhang, Y. Li, O. Liu, S. K. Kwok, W. H. Ip and O. Kaynak, An Efficient ant colony system based on receding horizon control for the aircraft arrival sequencing and scheduling problem, *IEEE Transactions on Intelligent Transportation System*, 2010, 11(2), 399–412.

7. W. N. Chen, J. Zhang, H. Chung, W. L. Zhong, W. G. Wu and Y. H. Shi, A novel set-based particle swarm optimization method for discrete optimization problems, *IEEE Transactions on Evolutionary Computation*, 2010, 14(2), 278–300.

8. Z. H. Zhan, J. Li, J. Cao, J. Zhang, H. Chung and Y. H. Shi, Multiple populations for multiple objectives: A coevolutionary technique for solving multiobjective optimization problems, *IEEE Transaction on System, Man, and Cybernetics B*, 2012, 43(2), 445–463.

9. Z. H. Zhan, J. Zhang, Y. H. Shi and H. L. Liu, A modified brain storm optimization, *Proceedings of the IEEE Congress on Evolutionary Computation*, 2012, pp. 1–8.

10. D. D. Zhou, Y. H. Shi and S. Cheng, Brain storm optimization algorithm with modified step-size and individual generation, In Y. Tan, Y. Shi and Z. Ji (Eds.), *Proceedings of the International Conference Swarm Intelligence*, Berlin, Germany, 2012, pp. 243–252.

11. J. Q. Xue, Y. Y. Wu, Y. H. Shi and S. Cheng, Brain storm optimization algorithm for multi-objective optimization problems, *Proceedings of the International Conference Swarm Intelligence*, Springer, Shenzhen, China, 2012, pp. 513–519.

12. H. Choset, K. Lynch, S. Hutchinson, G. Kantor, W. Burgard, L. Kavraki and S. Thrun, *Principle of Robot Motion: Theory, Algorithms, and Application*, MIT Press, Cambridge, MA, 2005.

13. E. Masehian and D. Sedighizadeh, Classic and heuristic approaches in robot motion planning—a chronological review, *Proceedings of the World Academy of Science, Engineering and Technology*, 2007, 23, 101–106.

14. R. Hassan, B. Cohanim and O. de Weck, *A Comparison of Particle Swarm Optimization and the Genetic Algorithm*, American Institute of Aeronautics and Astronautics, Austin, TX, 2004.

15. Q. Yuan-Qing, S. De-Bao, L. Ning and C. Yi-Gang, Path planning for mobile robot using the particle swarm optimization with mutation operator, *Proceedings of the International Conference on Machine Learning and Cybernetics*, IEEE, Shanghai, China, 2004, pp. 2473–2478.

16. W. Li, L. Yushu, D. Hongbin and X. Yuanqing, Obstacle-avoidance path planning for soccer robots using particle swarm optimization, *Proceedings of the IEEE International Conference on Robotics and Biomimetics (ROBIO)*, 2006, pp. 1233–1238.

17. C. Xin and L. Yangmin, Smooth path planning of a mobile robot using stochastic particle swarm optimization, *Proceedings of the IEEE on Mechatronics and Automation*, 2006, pp. 1722–1727.

18. L. Kavraki, P. Svestka, J.C. Latombe and M. Overmars, Probabilistic roadmaps for path planning in high-dimensional configuration spaces, *IEEE Transactions on Robotics Automation*, 1996, 12(4), pp. 566–580.

19. R. Eberhart and J. Kennedy, A new optimizer using particle swarm theory, *Proceedings of the Sixth International Symposium on Micro Machine and Human Science*, MHS, IEEE, Nagoya, Japan, 1995, pp. 39–43.

20. M. Dorigo, V. Maniezzo and A. Colorni, The ant system: Optimization by a colony of cooperating agents, *IEEE Transactions on Systems, Man, and Cybernetics Part B: Cybernetics*, 1996, 26(1), 1–13.

21. S. Kirkpatrick, C. D. Gelatt and M. P. Vecchi, Optimization by simulated annealing, *Science*, 1983, 220(4598), 671–680.

22. J. H. Holland, Adaptation in natural and artificial systems: An introductory analysis with applications to biology, Control and artificial intelligence. The University of Michigan Press, Cambridge, MA, 1975.

23. R. Storn and K. Price. Differential evolution—A simple and efficient heuristic for global optimization over continuous spaces, *Journal of Global Optimization*, 11, 341–359, 1997.

24. L. N. de Castro and F. J. Von Zuben. Artificial immune systems part I: Basic theory and applications. Universidade Estadual de Campinas, Dezembro de, *Tech. Rep*, 1999.

25. E. Alba, B. Dorronsoro, M. Giacobini and M. Tomassini, Decentralized Cellular Evolutionary Algorithms, *International Journal of Applied Mathematics and Computer Science*, 2004, 14(3), 101–117.

26. S. Roy, D. Banerjee, C. G. Majumder, A. Konar and R. Janarthanan, Dynamic obstacle avoidance in multi-robot motion planning using prediction principle in real environment, *Automation, Control and Intelligent Systems*, 2013, 1(2), 16–23, doi: 10.11648/j.acis.20130102.11.

27. A. Z. Nasrollahy and H. H. S. Javadi, Using particle swarm optimization for robot path planning in dynamic environments with moving obstacles and target, *Proceedings of the 3rd European Symposium on Computer Modeling and Simulation*, IEEE, Athens, Greece, 2009, pp. 60–65.

28. H. Gaber, S. Amin and M. Salem, A combined coordination technique for multi-agent path planning, *2010 10th International Conference on Intelligent Systems Design and Applications*, IEEE, Cairo, Egypt, 2010, 563–568.

29. L. Wang, L. Yong and M. Ang, Hybrid of global path planning and local navigation implemented on a mobile robot in indoor environment, *Proceedings of the 2002 IEEE International Symposium on Intelligent Control*, Vancouver, British Columbia, October 2002, pp. 821–826, doi: 10.1109/ISIC.2002.1157868.

30. Z. Zhan, W. Chen, Y. Lin, Y. Gong, Y.-L. Li and J. Zhang, *Parameter Investigation in Brain Storm Optimization*, IEEE Trans, 2013.

10

Intelligent Water Drop Algorithm

In this chapter, we will discuss intelligent water drop algorithm (IWDA or IWD)—a very successful algorithm in the field of optimization and discrete problems based on scheduling. The most important feature is the concrete *pseudo* communication between agents and the mathematics that represents them. Though this algorithm lacks parameter variation operation, yet it is one of the few algorithms, which is applicable to a wide variety of problems, including both continuous and discrete problems.

10.1 Intelligent Water Drop Algorithm

The nature-inspired computational family mainly comprises the algorithms, which are derived from nature or a natural phenomenon as discussed in Chapter 1. However, as all organisms and their surroundings belong to nature. It becomes really important to demarcate and subcategorize the algorithms and specify their classes. However, nature-inspired computation generally means all the nonorganic phenomenon of nature or rather those not related to any organism or its subsets as a whole. In this chapter, we discuss a very authentic nature-inspired algorithm that has been formulated directly from natural instincts of nature. The main specialty of the IWDA is that it is held by some nested chains of mathematical expressions and perhaps it is very difficult to visualize the dynamics of the particles or agents both in discrete and continuous domain of problems. However, as the discrete domain problems and graph-based problems demand, this algorithm provides an adequate opportunity for exploration and less for exploitation. When compared with the most trusted and pioneers of discrete-based algorithms such as ant colony optimization (ACO) algorithm, there are some major differences (ACO has been described in Chapter 4). In ACO algorithm, the working principle of the algorithm is dependent on the value of the two influencing factors, α and β, and these parameters decide the selection of the next movement. It is meaningless to mention that at any point of time, if the values of α and β become unity or $\alpha = \beta = 1$ then the algorithm will tend to become a greedy algorithm. In other words, the conditions and values of α and β make the algorithm a meta-heuristics one and help in extending the capability of the algorithm for better exploration as compared to exploitation. However, for IWDA all the agents tend to take the same edge due to the deterministic involvement of the parameters. However, if introduced with proper dynamic and self-adaptive parameters, this IWDA can also perform well and result in a better exploration and enhanced convergence for optimization.

10.1.1 Inspiration and Traditional Intelligent Water Drop Algorithm

IWDA was introduced by Hamed Shah-Hosseini in 2007 [1]. It is a "discrete graph-based specialized nature-inspired feeble-communicative and cooperative swarm-based meta-heuristic" algorithm. This algorithm is inspired by the path-determining capability of the moving water droplets (rivers, streams, etc.) and their interaction with the river bed under the influence of gravitational force. The algorithm is characterized by the ability of the water droplet to carry soil along with it through the course of its journey. The amount of soil that can be carried by a drop depends on its velocity factor. The water drops tend to follow a straight path under their natural instinct and influence unless there is any kind of obstruction or hindrance in the path [3,4]. As the accumulated water droplets move, they remain either under the influence of the *ease of the path* or the *opportunity*. One gets over the other to determine the course of the next action or the path to follow for the stream. Hence, the movement of water drops toward the best path does not always remain practical. The course of a discrete algorithm or meta-heuristic works well when it optimizes the path and does not depend on the deterministic exploitation of the graph-based network.

The algorithm is based on the observation that with an increase in the velocity of the water drop, there is a considerable change or rather an increase in the amount of soil carried by the water drop and this extra soil is derived out of the river bed.

Traditional Intelligent Water Drop Algorithm (TIWDA)

Step 1: Initialize the agent parameters, graph matrix, and other required storage structures.

For graph, each parameter of an edge is associated with a virtual soil value and is denoted as soil(i, j), which is read as the value of soil for the edge from node i to node j. Initially, all such soil values are same, but parameters differ and gradually they also change. There has been an investigation (for the modified IWDA) that whether for each edge there must be a separate soil value for each parameter or a combined one for all of them. In addition, for the agents, there is a virtual velocity initialized to some value and is given as V_k^{IWD} for the kth agent or water drop.

Step 2: For each particle agent k at city i having m options to select node j from the selection depends on the calculation of the probability P_{ij}^{IWD} given as

$$P_{ij}^{IWD} = \frac{f(\text{soil}(i,j))}{\sum_{k \in S, k \neq i} f(\text{soil}(i,k))} \qquad (10.1)$$

Where $f(\text{soil}(i, j))$ is defined as

$$f(\text{soil}(i,j)) = \frac{1}{\epsilon_s + g(\text{soil}(i,j))} \qquad (10.2)$$

And $g(\text{soil}(i, j))$ is defined as

$$g(\text{soil}(i,j)) = \begin{cases} \text{soil}(i,j) & \text{if } \dfrac{\min}{l \notin V_c(\text{IWD})}\, \text{soil}(i,l) \geq 0 \\[2ex] \text{soil}(i,j) - \dfrac{\min}{lV_c(\text{IWD})}\, \text{soil}(i,l) & \text{otherwise} \end{cases} \qquad (10.3)$$

Here, ϵ_s is a small factor to prevent the denominator from becoming zero. Now, we cannot say anything about the path selected because the parameters of the path or edges are not involved; but the fact remains that all the particles will come up with the same path unless the virtual parameters have been changed. We have introduced a random parameter to perform more exploration in the network and have explained it in detail in the next subsection (10.3).

Step 3: Now, the movement of the agents will induce some change in the virtual velocity and sand content of the individuals but the total sand content will remain the same for the system consisting of the agents and the graph network. For an agent that is moved from node i to node j, the updated velocity is given by

$$vel^{IWD}(t+1) = vel^{IWD}(t) + \frac{a_v}{b_v + c_v * \text{soil}(i,j)} \tag{10.4}$$

Here, a_v, b_v, and c_v are constants.

Step 4: Now, the change in the soil content of the water drop of the agent is given by $\Delta\text{soil}(i,j)$ for its passage from node i to node j as

$$\Delta\text{soil}(i,j) = \frac{a_s}{b_s + c_s * \text{time}(i,j;vel^{IWD}(t+1))} \tag{10.5}$$

Here, we have defined $\text{time}(i,j;vel^{IWD}(t+1))$ as

$$\text{time}(i,j;vel^{IWD}(t+1)) = \frac{\text{HUD}(j)}{vel^{IWD}(t+1)} \tag{10.6}$$

Here, $\text{HUD}(j)$ is some calculated factor derived out of the real parameter values and depends on the representation of the problem in the graph.

Step 5: Now, the change in the content of the soil for the edge between node i and node j is given by the following equation:

$$\text{soil}(i,j) = (1 - \rho_n) * \text{soil}(i,j) - \rho_n * \Delta\text{soil}(i,j) \tag{10.7}$$

and the update for water drop is given by

$$\text{soil}^{IWD} = \text{soil}^{IWD} + \Delta\text{soil}(i,j) \tag{10.8}$$

Step 6: Now, the soil level of the whole passage of the optimized path is further updated as

$$\text{soil}(i,j) = (1 + \rho\text{IWD}) * \text{soil}(i,j) - \rho\text{IWD} * \frac{1}{(N_{IB} - 1)} * \text{soil}_{IB}^{IWD} \forall (i,j) \in T_{IB}, \tag{10.9}$$

where we have defined T_{IB} as $T_{IB} = \arg_{\forall T^{IWD}} \overset{max}{} q(T^{IWD})$

10.2 Intelligent Water Drop Algorithm for Discrete Applications

The algorithm presented previously is described by the following pseudocode and flowchart. This will surely give a better understanding of the algorithm, various modifications possible, pros, and cons of each such modification as well as hybridization issues. It needs to be remembered that as we are talking of a system that is dynamic in nature, we must not expect a well-defined value for the global optimum. The notion of global optimum is valid however, for an unchanged state or specifically for each state of parameters.

Another pressing issue is the selection of stopping criteria for the search process and in most of the cases, this selection will depend on the application concerned. Some researchers have endorsed the idea of *full utilization* in which the stopping criterion is met only when all drops have reached their destination. This approach brings with itself, looping in the network unless some counter measures are also taken. Another choice for the stopping criteria could be the maximum number of nodes that are present in the graph; even with this choice, there is a chance that majority of the drops may not have reached their destination. Hence, the best path may not appear altogether. As far as the *best solution foundation* is concerned, the first selection metrics involves more number of iterations for the algorithm to converge. With load balancing, the second selection metrics fairs well as the requirement for a successful load balancing may remain independent of the best path. Instead of the best path, this may concern itself with a different or an underutilized path (that might have cropped up due to the forced end of the iteration). The fitness function for the multi-objective problem is chosen to be a nonweighted multiparameter function given by

$$\text{fitness} = (f_1 + f_2) \tag{10.10}$$

10.2.1 Intelligent Water Drop Algorithm for an Optimized Route Search

Step 1: $G = (V, E)$, matrix representing road and matrix containing parameters for each edge of the graph are initialized.

Step 2: N intelligent water droplets, path traverse vector for each drop, number of nodes present in the graph, destination-reached factor, sum of parameter vector, fitness $\{f_1, f_2, ..., f_n\}$ are initialized.

Step 3: Initial sand content for each sand type $\{s_1, s_2, ..., s_m\}$ and for each edge of the network is initialized.

Step 4: Drop starts from the source node.

Step 5: Drop is moved through the graph following the probabilistic equations of minimal sand level.

Step 6: Velocity of the drop is updated.

Step 7: "Time Spent Equivalent" for the IWD swarms is determined at their respective edges.

Step 8: Sand content of the drop is updated.

Step 9: Sand content of the river bed is updated.

Step 10: The parameters—path traversed vector, number of nodes, destination-reached factor (If reached), sum of parameter vector for each IWD are updated.

Step 11: While stopping criterion is not reached

Go to Step 5 and Continue from there

Step 12: Fitness of each IWD is estimated.

Step 13: Best IWD is estimated during each iteration.

Step 14: Best IWD path is set as "Best Path Vector"

Step 15: Information is provided to the vehicles.

Step 16: If (change in Dynamic Parameter is greater than Threshold) OR (Certain Time Limit has elapsed) then

Restart Iteration from Step 2

The graphs shown below represent the performance of different variants of IWD algorithm. In these graphs, the number of iterations and the time spent in the run of each algorithm is considered (Figures 10.1 through 10.4) [6]. The variants discussed are IWD1 (SC1), IWD2 (SC2), IWD3 (SC3 with sand denoting only travel time), and IWD4 (SC3 with sand denoting only waiting time).

As can be seen very easily, IWD3 has a better convergence rate than IWD4 (see Figure 10.4), whereas in Figure 10.2, IWD4 outperforms IWD3. The convergence rate

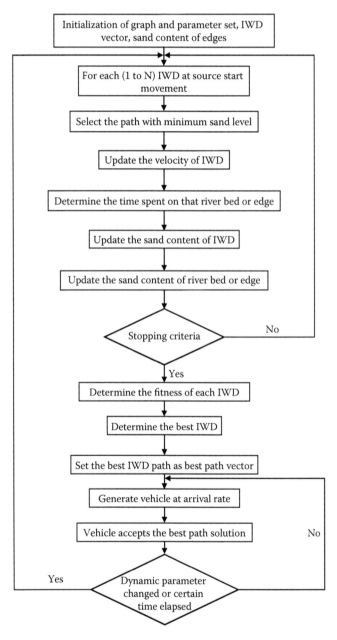

FIGURE 10.1
Flow diagram for the IWDA implemented.

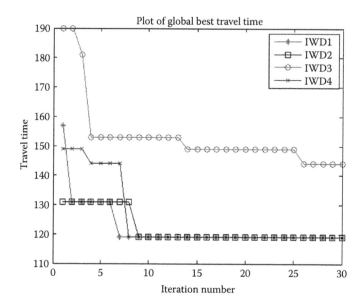

FIGURE 10.2
Plot for the comparison of global best with respect to travel time (f_1).

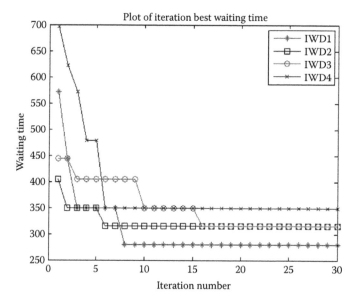

FIGURE 10.3
Plot for the comparison of global best with respect to waiting time (f_2).

of IWD1 is above par compared to other variants. This can be attributed to the fact that in SC1, both the parameters create changes on the same sand and consequently the *change in parameter* that actually happens to dominate the equations provides for better results as compared to other variants. The choice of functions f_1 and f_2 remains of paramount importance and in the light of this, we have chosen a linear function.

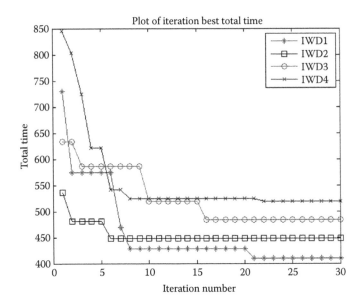

FIGURE 10.4
Plot for the comparison of global best with respect to total time $(f_1 + f_2)$.

The primary reason for considering such a linear model is that when dealing with such functions, we can log the sand values, particularly when the same sand is considered for all the parameters. For simple path planning algorithms, undesirability parameter can be interpreted as a distance, but for multiobjective optimization it has to remain a function of the parameters.

The advantage that we get using a selectively adaptive version of the algorithm is the addition of exploration as well as exploitation faculties. This ensures that none of the tributaries and distributaries remain dry. In addition, the selection of alternative routes will bring forth new and opportunistic node combinations and at the same time will help the algorithm to procure global optimization. The probability terms present in the update equations will also help in the initial stages (when most of the edges are still virgins and there is no clue for determination of path probability). The effectiveness of the algorithm can be considered for different combinations of network dynamics and parameter values. In the above-mentioned case, the adaptive feature of the algorithm will aid in the determination of a better path.

10.2.2 Intelligent Water Drop Algorithm Convergence and Exploration

IWDA is mainly suitable for graph-based problems in which it takes care of two aspects—one is the parameters it generates for search decision-making and the other is the parameters of the search space related to the application. Although convergence is mainly related to optimization of mathematical equation-based optimizations, exploration is related to both discrete and continuous search. Similarly for an equation, the optimized value of which is to be searched, IWDA will derive the combination of the variables in the search space. It can be a D-dimensional problem in which each solution can be of D dimensions. In this case, each position of the search space is a solution. For instance, in a 2D search space the point with minima or maxima is the solution as IWDA is highly explorative and least dependent on the local aspects or the information it can explore. With a change in the number of IWD agents, the convergence will vary. However, as the exploration is mainly dominant, the IWDA will converge to the required solution. But in case of a graph-based or combinatorial problem, the solution gradually grows and is not confined to just some fixed D dimensions [5]. It can vary with the problem and the objective statement. The main criteria governing such problems is the size of the search space or more accurately, the state space. So, in that case, a combination of states, satisfying certain constraints, will be the solution and the best solution with respect to some evaluation criteria will be regarded as the final solution. IWDA has that capability to explore various combinations of state spaces and come up with a solution. However, it must be mentioned here that the IWDA works on a generalized situation and this is the reason why it has been accepted on various applications. As various problems have diverse constraints and bounds, each of such cases must be handled individually and care must be taken that the solutions generated are not violating them.

10.3 Variants of Intelligent Water Drop Algorithm

10.3.1 Adaptive Intelligent Water Drop Algorithm

To introduce more involvement and better exploration, there are several changes that are being made on the classical IWDA. These changes are described as the modified intelligent water drop algorithm (MIWDA). These adaptive parameters introduce better dynamics for the agents in the system. The changes mainly involve making the parameters and numerical variables probabilistic and adaptive so that some fluctuation occurs in the decision-making of the algorithm. The algorithm described here is with respect to optimization of a road network and the main characteristics of the analysis are motivated with multiobjective optimization in which they reoccur in more than one parameter that is attached with each edge of the graph network. Hence, the example equation and implications attached with the road network cases are included with the main track of the MIWDA.

The MIWDA is not divided into steps due to the inclusion of a road network case demonstration side by side, but the step guidance found in the IWDA can be extended with this MIWDA.

In the modified algorithm, the probability of movement of water drop from node i to node j is computed [6] as

$$p^{\text{IWD}}(i,j) = \frac{\alpha_{ij} * f(\text{soil}(i,j))}{\sum_{k \notin V_c(\text{IWD})} f(\text{soil}(i,k))} \tag{10.11}$$

Here, α_{ij} is the number chosen randomly from $[-1,1]$, $f()$ represents the soil content in the path (equivalently, an edge) and k is a node chosen such that it is not the node from where the flux is coming from. The introduction of α_{ij} parameter introduces better exploration capabilities than mere exploitation. When the minimum distance is chosen as a metric to get the path, the algorithm ends up giving the same results in each iteration. As a result, the iteration best and the global best solutions coincide. However, for a system that is dynamic, the waiting time of the edges keeps on changing. Hence, the concept of path planning on an iteration-wise basis remains short-lived; this demands the iteration to incorporate more exploration capabilities. The α_{ij} factor, when used with a negative sign, may consider the worst path as the best as an optimal routing experience.

At this stage, $f()$ can be defined as

$$f(\text{soil}(i,j)) = \frac{1}{\epsilon_s + g(\text{soil}(i,j))} \tag{10.12}$$

In the above-mentioned equation, ϵ_s is a small constant; it ensures that the denominator of the right-hand side does not become zero. Following this, we define $g(\text{soil}(i,j))$ function as

$$g(\text{soil}(i,j)) = \begin{cases} \text{soil}(i,j) & \text{if} \dfrac{\min}{l \notin V_c(\text{IWD})} \text{soil}(i,l) \geq 0 \\[2ex] \text{soil}(i,j) - \dfrac{\min}{lV_c(\text{IWD})}\text{soil}(i,l) & \text{otherwise} \end{cases} \tag{10.13}$$

The condition given in the above-mentioned equation signifies that the value of $g(\text{soil}(i,j))$ must decrease at any cost, and hence the condition. If the value of $g(\text{soil}(i,j))$ increases, then it is by default the best path to pursue.

The velocity of the drop remains path dependent and the updated velocity is given by

$$vel^{\text{IWD}}(t+1) = vel^{\text{IWD}}(t) + \frac{a_v(t)}{b_v(t) + c_v(t)*\text{soil}^2(i,j)} \tag{10.14}$$

Now, the soil content of both the IWD and the river bed increases or decreases by the same amount and the change $\Delta\text{soil}(i,j)$ is given by

$$\Delta\text{soil}(i,j) = \frac{a_s(t)}{b_{s(t)} + c_s(t)*\text{time}^2(i,j;vel^{\text{IWD}}(t+1))} \tag{10.15}$$

As can be seen, for a given edge, the change in soil content varies inversely with the time spent by the IWD in that edge.

10.3.2 Same Sand for Both Parameters (SC1)

In this variant, the optimization criteria chosen are the waiting time and distance. These parameters can get recessive; this is attributed to the fact that there is a tendency of the shortest route to get congested. This is because the initial solution will compute the path with the minimum distance and the least time. However, after few iterations, the algorithm will try to get the best optimal route. It can be asserted because; following a rush of vehicles, the waiting times will gradually increase. In this case, we have assumed that only one type of sand will be involved in the algorithm for optimization and decision-making and that the factors such as velocity change, sand content of droplets, sand content of paths, and so on will remain dependent on it. In addition, the time spent in the paths will be a summation of travel time (*TT*) as well as waiting time (*WT*).

$$\text{time}(i, j; vel^{\text{IWD}}(t+1)) = TT + WT \tag{10.16}$$

$$TT = \frac{\text{HUD}(j)}{vel^{\text{IWD}}(t+1)} \tag{10.17}$$

where WT is a dynamic parameter equal to the average waiting time at the crossings.

10.3.3 Different Sand for Parameters Same Intelligent Water Drop Can Carry Both (SC2)

In this variant, decision-making scheme, velocity change, and sand content change are controlled by both the sand types instead of just one type. The modified equations [6] are

$$p_{\text{total}}^{\text{IWD}}(i, j) = \prod_{\forall x} \frac{\alpha_{ij} * f(\text{soil}_x(i, j))}{\sum_{k \notin V_c(\text{IWD})} f(\text{soil}_x(i, k))} \tag{10.18}$$

$$f(\text{soil}(i, j)) = \frac{1}{\epsilon_s + g(\text{soil}(i, j))} \tag{10.19}$$

$$g(\text{soil}_x(i, j)) = \begin{cases} \text{soil}_x(i, j) & \text{if } \min_{l \notin V_c(\text{IWD})} \text{soil}_x(i, l) \geq 0 \\ \text{soil}_x(i, j) - \min_{l V_c(\text{IWD})} \text{soil}_x(i, l) & \text{otherwise} \end{cases} \tag{10.20}$$

$$vel^{\text{IWD}}(t+1) = vel^{\text{IWD}}(t) + \frac{a_v(t)}{b_v(t) + c_v(t) * \text{soil}^2(i, j)} \tag{10.21}$$

$$a_v(t) = a_v(t-1) + \text{constant}_{a_v} * \text{rand}() \tag{10.22}$$

$$b_v(t) = b_v(t-1) + \text{constant}_{b_v} * \text{rand}() \tag{10.23}$$

$$c_v(t) = c_v(t-1) + \text{constant}_{c_v} * \text{rand}() \tag{10.24}$$

$$\Delta \text{soil}_x(i, j) = \frac{a_s(t)}{b_{s(t)} + c_s(t) * \text{time}_x^2(i, j; vel^{\text{IWD}}(t+1))} \tag{10.25}$$

$$\text{time}(i, j; vel^{\text{IWD}}(t+1)) = \begin{cases} TT & \text{First Case} \\ WT & \text{Second Case} \end{cases} \tag{10.26}$$

$$\text{soil}_x(i, j) = (1 - \rho_{nx}) * \text{soil}_x(i, j) - \rho_{nx} * \Delta \text{soil}_x(i, j) \tag{10.27}$$

$$\text{soil}_x^{\text{IWD}} = \text{soil}_x^{\text{IWD}} + \Delta\text{soil}_x(i,j) \tag{10.28}$$

Now, for a given iteration and parameter set, there occurs an iteration best solution. This solution is given by

$$T^{IB} = \arg{}_{\forall T^{\text{IWD}}}\max q(T^{\text{IWD}}) \tag{10.29}$$

where $q()$ is called the *quality function*.

10.3.4 Different Sand for Parameters Same Intelligent Water Drop Cannot Carry Both (SC3)

This variant, similar to the one discussed earlier, considers different sand for parameters; however, the restriction is that one water droplet type may carry only one type. Thus, there will be two types of droplets each having different sand types. These droplets will simultaneously (also independently) search the same data space and look for solutions. Using the usual notations, this variant can be modeled by the following equations:

$$p_{\text{total}}^{\text{IWD}}(i,j) = \prod_{\forall x} \frac{\alpha_{ij} * f(\text{soil}_x(i,j))}{\sum_{k \notin V_c(\text{IWD})} f(\text{soil}_x(i,k))} \tag{10.30}$$

$$f(\text{soil}(i,j)) = \frac{1}{\epsilon_s + g(\text{soil}(i,j))} \tag{10.31}$$

$$g(\text{soil}(i,j)) = \begin{cases} \text{soil}(i,j) & if \quad \min_{l \notin V_c(\text{IWD})} \text{soil}(i,l) \geq 0 \\ \text{soil}(i,j) - \min_{l V_c(\text{IWD})} \text{soil}(i,l) & otherwise \end{cases} \tag{10.32}$$

Here, the velocity of water drops is path dependent. As such, the updated velocity is computed as

$$vel^{\text{IWD}}(t+1) = vel^{\text{IWD}}(t) + \frac{a_v(t)}{b_v(t) + c_v(t) * \text{soil}^2(i,j)} \tag{10.33}$$

$$\Delta\text{soil}_x(i,j) = \frac{a_s(t)}{b_{s(t)} + c_s(t) * \text{time}_x^2(i,j;vel^{\text{IWD}}(t+1))} \tag{10.34}$$

$$\text{time}(i,j;vel^{\text{IWD}}(t+1)) = \begin{cases} TT & \text{First Case} \\ WT & \text{Second Case} \end{cases} \tag{10.35}$$

$$\text{soil}_x(i,j) = (1 - \rho_{nx}) * \text{soil}_x(i,j) - \rho_{nx} * \Delta\text{soil}_x(i,j) \tag{10.36}$$

$$\text{soil}_x(i,j) = (1 + \rho\text{IWD}) * \text{soil}_x(i,j) - \rho\text{IWD} * \frac{1}{(N_{IB} - 1)} * \text{soil}_{IB}^{\text{IWD}} \forall(i,j) \in T_{IB} \& \forall x \tag{10.37}$$

10.4 Scope of Intelligent Water Drop Algorithm for Numerical Analysis

The scope of this algorithm has been so far confined to discrete problem such as the event-based or node-based problems. However, it can be used for other numerical analysis and optimization problems as well (assuming that the search space is transformed into some matrix and there are hundreds and thousands of numerical nodes with multidimensional parameters). One thing to keep in mind is that there are situations when the algorithm may not perform well as far as convergence and local search are concerned. The reasons for this limitation are its more explorative attitude and less adaptive approach for local information. Exploitation is good only when we are optimistic but the real search occurs only when the algorithm behaves differently and out of its comfort zone. To summarize, it can be said that the numerical analysis of the algorithm will improve when it is combined with some other algorithm. The combining algorithm in a way complements the base algorithm.

10.5 Intelligent Water Drop Algorithm Exploration and Deterministic Randomness

IWDA is an exploration-based algorithm that makes use of deterministic parameters. These features are either available in the search space or can be obtained from the local heuristic information. This makes the algorithm very dynamic with varying random options. This deterministic approach also helps in exploitation; however, the introduction of the factor α in the probability generation equation makes the water drop agents move in different directions and as a result, creates better combinations and varied options. The agents sometimes tend to go astray to some unavoidable paths, but it is part of the search; and for these cases, the number of agents must be carefully initialized, so as to achieve a balance between the required computational time and energy.

10.6 Related Applications

Since this algorithm mimics the behavior of a water drop, it can be directly applied in any engineering application pertaining to agriculture. The sole motivation remains that the mapping of abstract entities with the real-world entities will be flawless and direct. One such application is discussed in [7]. Here, growth of a Sunagoke moss is discussed. One of the foremost requirement for the growth of this plant is the availability of water and that too in moderation. Water levels not in moderation cause *water stress* within the plant; this stress is then captured, measured, and monitored by imaging techniques. In this work, authors have used IWDA to find the most significant set of textural features (TFs) for proper growth of the plant. These sets are usually obtained from gray, RGB, HSV color spaces, and have dimensionality of 120. When the results of neural IWD were compared with other feature selection (FS) methods, the former proved to be very efficient (Figure 10.5) [7]. The experimental procedure is described in the following flowchart:

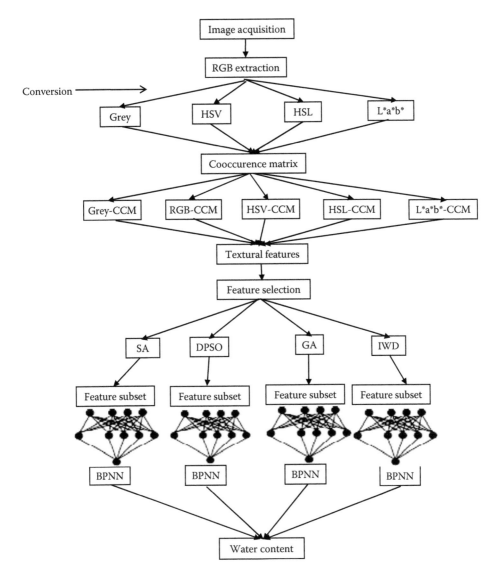

FIGURE 10.5
Experimental design.

The results of the experiment have been put in the following graphs (Figures 10.6 and 10.7) [7].

Another major field in which this algorithm has been used is power systems. The reactive power dispatch problems are quite common in power systems; the issue gets serious when they are trying to optimize multiple objectives at the same time. In [8], this problem has been solved using the IWD algorithm. Modal analysis of the system has been done to get the static voltage stability assessment. The objective was minimization and maximized voltage stability margin.

This algorithm is also used to solve known NP-Hard problems of computer science such as the knapsack problem [9]. For this, the original algorithm is modified so that the new algorithm includes a suitable local heuristics for multiple knapsack problem (MKP).

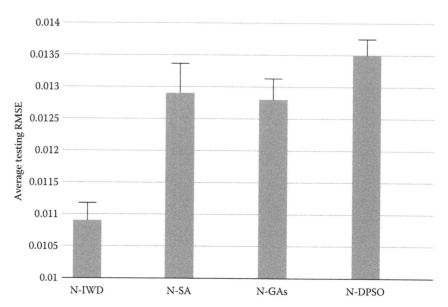

FIGURE 10.6
Comparison of N-IWD, N-SA, N-Gas, and N-DPSO based on average testing-set RMSE.

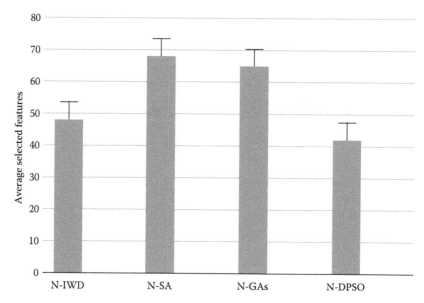

FIGURE 10.7
Comparison of N-IWD, N-SA, N-GAs and N-DPSO based on average number of selected relevant features.

Following this, the modified algorithm is used to solve this MKP. The proposed algorithm is tested by standard problems and the results demonstrate that IWD–MKP [2] algorithm is a promising alternative to solving such problems.

The experimental conditions and the results are described in Table 10.1 [9].

IWD has also been used in mobile ad hoc wireless networks (Figure 10.8) [9]. In these networks, the nodes remain independent, self-managed, and at the same time highly

TABLE 10.1

OR Library Problems That Are Solved by IWD–MKP Algorithm

Constraints × Variables	Quality of Optimum Solution	The Solution Quality of the IWD–MKP	Average Number of Iterations of IWD–MKP
10×6	3,800	3,800	3.3
10×10	8,706.1	8,706.1	12.9
10×15	4,015	4,015	30.9
10×20	6,120	6,120	18.7
10×28	12,400	12,400	11.9
5×39	10,618	10,563.6	100
5×50	16,537	16,405	100

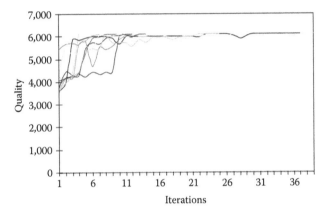

FIGURE 10.8
Convergence curves for the first 10 runs of IWD–MKP problem.

dynamic. [10] introduces a novel QoS aware multipath routing algorithm IWDRA; this is based on the IWD algorithm in which the data packets mimic the behavior of water drops. Experimental results have shown that when data packets mimic the behavior of IWD, the performance of the network improves. In order to quantify the performance of the network, parameters such as network lifetime, network stability, and packet delivery rate, and so on have been considered.

10.7 Summary

In this chapter, IWDA has been discussed in detail and examples of its implementations for any graph-based search space have also been covered. To summarize, the salient features of the algorithm are as follows-the success of IWDA lies in its ability to perform a heuristic search through a graph-based problem or any other search space that can be considered as a replica of the combinatorial optimization problems through manipulation of its own generated parameters and through other local heuristic application-specific information. Complex equations sometimes make the IWDA computationally heavy but it is worth it as it incorporates a better exploration. The features mentioned earlier also prevent the algorithm from becoming a greedy algorithm. As a result, the probability that the algorithm will hit a better solution is very high.

Solved Question

1. Highlight an application of IWDA algorithm stating clearly the modifications done to the original algorithm to realize the application.

Solution

A number of variants for IWDA are possible and they are chosen as per the requirements of the application. One such variation involves simplifying the equations of IWDA (to get results in a fewer iterations). Clearly, such an approach is feasible only when we can tolerate some error in our solutions and in which the solutions are needed within a scheduled deadline. One such simplification is discussed in [11]. Here, a simplified version of traditional IWDA is simplified and then run in a *computationally limited environment*. Further, the results were also compared with other algorithms. Results confirmed that if given a value for tolerance, simplified algorithm was able to give the result in a fewer iterations.

Unsolved Questions

1. IWDA is a "feeble-communicative and cooperative swarm-based meta-heuristic algorithm." Explain the *feeble-communicative* nature of the algorithm.
2. How is a balance achieved between exploration and exploitation in a IWDA algorithm?
3. Comment on the inherent weaknesses of IWDA algorithms.
4. In what circumstances does IWDA perform better than other contemporary algorithms?
5. What is the physical significance of heuristic undesirability HUD ()?
6. How is the effect of gravitational force implemented in an IWDA algorithm?
7. What is meant by a *heuristic search*? Comment on its usefulness.
8. Does the quality function affect the convergence of the algorithm? Explain.
9. Why is there a need to update the soil content of the river bed?
10. How does the representation of data affect the performance of IWDA algorithm?

References

1. S. H. Hamed, Problem solving by intelligent water drops, *Proceedings of the IEEE Congress on Evolutionary Computation*, Singapore, 2007, pp. 3226–3231.

2. S. H. Hamed, Intelligent water drops algorithm: a new optimization method for solving the multiple knapsack problem, *International Journal of Intelligent Computing and Cybernetics*, 2008, 1(2), 193–212.

3. S. H. Hamed, The intelligent water drops algorithm: A nature-inspired swarm-based optimization algorithm, *International Journal of Bio-Inspired Computation*, 2008, 1(1), 71–79.

4. H. S. Hosseini, Optimization with the nature inspired intelligent water drops algorithm, *International Journal of Intelligent Computing and Cybernetics*, 2008, 1(2), 193–212.

5. C. Blum and A. Roli, Metaheuristics in combinatorial optimization: Overview and conceptual comparison, *ACM Computing Survey*, 2003, 35(3), 268–308.

6. C. Sur, S. Sharma and A. Shukla, Multi-objective adaptive intelligent water drops algorithm for optimization & vehicle guidance in road graph network, *2nd International Conference on Informatics, Electronics & Vision*, 2013, pp. 17–18.

7. Y. Hendrawan and H. Murase, Neural-intelligent water drops algorithm to select relevant textural features for developing precision irrigation system using machine vision, *Computers and Electronics in Agriculture*, 2011, 77(2), 214–228.

8. K. Lenin and M. S. Kalavathi, An intelligent water drop algorithm for solving optimal reactive power dispatch problem, *International Journal on Electrical Engineering and Informatics*, 2012, 4(3).

9. H. S. Hosseini, Intelligent water drops algorithm: A new optimization method for solving the multiple knapsack problem, *International Journal of Intelligent Computing and Cybernetics*, 2008, 1(2), 193–212.

10. D. Sensarma and K. Majumder, IWDRA: An intelligent water drop based qoS-aware routing algorithm for MANETs, *Proceedings of the International Conferences on Frontiers of Intelligent Computing: Theory and Applications*, 2013, 247.

11. J. Straub and E. Kim, Characterization of extended and simplified intelligent water drop (SIWD) approaches and their comparison to the intelligent water drop (IWD) approach, *25th International Conference on Tools with Artificial Intelligence*, 2013, pp. 101–107.

11

Egyptian Vulture Algorithm

11.1 Introduction

The Egyptian vulture optimization algorithm (EVOA) is a meta-heuristic algorithm that was acquainted principally to tackle the combinatorial issues. The process of obtaining the sustenance by Egyptian vulture created this algorithm. Hard optimization problems can be solved if we use the wise conduct of this creature as an algorithm. These versatile, creative demonstrations of Egyptian vulture make it as a standout among the bird species.

This meta-heuristics can be applied for global solutions of the combinatorial optimization problems [1] and has been considered on the customary 0/1 knapsack problem (KSP) and tried for a few datasets of various measurements. The algorithm performed very well when it was applied on problems such as KSP and traveling salesman problem (TSP) [2] and the results obtained were close to an ideal value and give the extent of use in comparable problems such as path planning and various combinatorial optimization problems.

The essential nourishment of Egyptian vulture is flesh; they eat the eggs of other birds instead of eating flesh of other creatures. The pebbles are tossed as mallet to break the eggs of other birds. Rolling with twigs is another intriguing feature of this bird. They are recognized from other birds because of their ability of rolling objects with twigs.

The two fundamental activities portrayed earlier are transformed into an algorithm termed as EVOA.

11.2 Motivation

Various real-life situations are represented by graph-based NP-hard problems and constrained discrete domain problems and it is very hard to accomplish these problems using algorithms and mathematical operators. There is a similar problem with mathematically critical issues (such as sequential ordering problems) and combinatorial optimization problems such as path planning. These sorts of issues need algorithms, which are converging, randomized, and fit for generating combinations efficiently and can hit the near optimized solution after significant number of iteration. With such a variety of nature-inspired heuristics for optimization such as particle swarm optimization (PSO) [3], honey bee swarm [4], league championship algorithm [5], cuckoo search (CS) by means of levy flights [6], bat algorithm (BA) [7], simulated annealing [8], differential evolution [9], harmony search [10], glowworm swarm optimization [11], Honey bee mating optimization (HBMO) algorithm [12], Krill herd algorithm, virus optimization algorithm [19], and so on. Many of them have been described in previous chapters and some of them will be described in the upcoming chapters. The vast majority of them are not good for problems having discrete states but these algorithms have good performance in continuous domain problems and are more suitable for exploration of variation of parameter-based search problems. The bio-inspired algorithms depicted in [2–19] are mostly competent for continuous domain combination and if at any point they are compelled to be used on the discrete space issues, one issue exists, i.e., once a series or sequence is shaped either the entire algorithm is modified or a confined part of it is modified, yet they do not have the operator for inferring the change of sequence in between.

In this chapter, another bio-inspired meta-heuristic EVOA is given, which is promptly appropriate for the graphical issues and node-based continuity search and optimization ability and there is no necessity of local search in the continuous space. The principle preferred in the standpoint of the EVOA is its ability for various combination formations and in the meantime prevents loop formation furthermore in numerous issue case (such as TSP and Knapsack) addition and consumption of node sequence, voluntarily for the improvement of the solution and in the meantime generation of new arrangement sequence.

11.3 History and Life Style of Egyptian Vulture

The Egyptian vulture, is also called the white scavenger vulture and its biological name is *Neophron percnopterus*. It is a standout among the most antiquated sorts of vulture that lived on this Earth and it has features similar to dinosaurs family concerned with its sustenance habit, quality, perseverance, and in intelligence it has outperformed them; however, shockingly a couple of types of its kind has turned out to be extinct. Similar to some other vulture species, the essential nourishment living space of the Egyptian vulture is meat; however, the sharp element for sustenance habit, which makes it extraordinary in consuming the eggs of other birds it leads to a new meta-heuristic. Figure 11.1 [2] demonstrates the Egyptian vulture in the procedure of eating eggs of different birds.

FIGURE 11.1
Egyptian vulture algorithm.

In any case, pebbles are tossed very hard on bigger and solid (as far as fragility) eggs; stones are used as hammer and by this they break the eggs. Rolling things with twigs is the skill of Egyptian vulture, which is another recognizing feature of this bird. Apart from chasing for sustenance, the life of this bird is inactive; the thing that makes them best among all the individuals in the class of Aves is their level of execution and remarkable strategy. However, because of some unavoidable reasons of the deceitful action of the individuals such as poaching, chopping down of woods, an unnatural weather change, and so forth, there has been a diminishing in their numbers in the populace.

This bird was famous as an image of eminence and was termed as *Pharaoh's Chicken* and has been celebrated among the Pharaohs of Egypt. There are superstitious believes connected with the bird, in India. Downfall, hardship, or even death are considered as the result of shrilling calls and even seeing this bird is considered as unfortunate. Natural catastrophe or occurrence can be identified by some tribal groups and these groups really tame them. There is another superstition that is related to these birds. There is a temple in Thirukalukundram in Chengalpattu, India and it is famous for a couple of Egyptian vulture, which visits the place for a considerable length of time. These vultures used to arrive before noon to eat on the offering produced using the customary Indian nourishment such as wheat, sugar, ghee, and rice, and were sustained ritualistically by the priests of the temple. When there is any delay in the arrival of the Egyptian vulture, it is credited to the nearness of any *miscreants* among the spectators but usually they are prompt in their arrival. Likewise, as per mythology legend 8 sages are represented by the vultures (or *hawks*) who were rebuffed by Lord Shiva (Hindu God), and two of them leave with the progression of ages.

11.4 Egyptian Vulture Optimization Algorithm

The EVOA meta-heuristics algorithm has been depicted here in steps that are represented through illustrations and clarifications. The capacity of moving things with twigs and tossing of stones are two fundamental exercises of the Egyptian vulture and these steps are transformed into algorithm and that algorithm is known as EVOA (Figure 11.2) [2].

These are the steps of general EVOA and due to the constraints of the problem some of the stages can be twisted or restricted for adaptation.

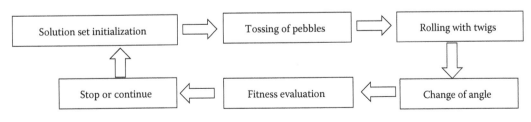

FIGURE 11.2
Steps for Egyptian vulture algorithm.

Step 1: Solution set or strings are initialized, which have the representations of parameters and it is in the form of variables. A single state of allowable solution is represented by the string that shows a set of parameters.

Step 2: Representative variables are refined, superimposed constraints and conditions are checked.

Step 3: Stones are tossed at selected or random points.

Step 4: Rolling of twigs operation is performed on an entire string or selected part of it.

Step 5: Selective part of the solution is reversed using change of angle.

Step 6: Evaluation of fitness.

Step 7: Stopping criteria is checked.

A sequence of serially associated events or states is built up, which respects a few constraints and this solution is optimized with some parameter(s) as a whole or independently, and this is the fundamental objective of the algorithm, which is achieved after the end of simulation.

11.4.1 Pebble Tossing

The vulture breaks the harder eggs of another bird and gets the nourishment from inside by utilizing the stones for breaking it. A few Egyptian vultures attempt to find feeble points or splits and forcefully they toss the stones on the eggs in a constant manner and try to break it. New solutions are introduced randomly at specific positions in the solution set using this approach as a part of this meta-heuristics and for the selection of extension of the performance and execution of the operators, two parameters are produced and based on the probability set it may achieve four potential outcomes and all this is possible because of breaking of the solution. Pictorial representation of the pebble tossing step is shown in Figure 11.3 [2]. Here, nodes or discrete events are represented by numerical values of a solution string or array and parameters are represented by every component of the array. All these things are valid for Figures 11.3 through 11.5. Only one of the cases will happen but all these three outcomes are possible. For the purpose of clarification, all three possibilities are shown.

In order to determine the extent of operation the two variables are used:

PS is the Pebble Size (occupancy level) where $PS \geq 0$

FT is the Force of Tossing (removal level) where $FT \geq 0$

Hence, If $PS > 0$ then "Get In" Else "No Get In."

Also, if $FT > 0$ Then "Removal" else "No Removal."

For removing there is "Removal" and occupancy is denoted by "Get In." How many solutions should a pebble contain in the solution set that must be introduced forcefully and it is denoted by the level of occupancy. The level of removal implies the number of solutions that should be removed.

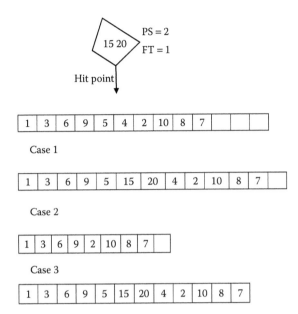

FIGURE 11.3
Pictorial view of pebble tossing.

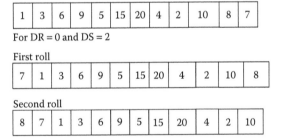

FIGURE 11.4
Pictorial view of rolling with twigs for DS = 2.

| 1 | 2 | 3 | 4 | 7 | 6 | 5 | 8 | 9 | 10 | | | |

Say (1,2,3,4) forms a link, (7,6,5) another link, (8,9,10) another one. But there is no link between 4,7 and 5,8. But change of angle reverses the link 7,6,5 and tries to see if link exists between 4,5 or 7,8 or both.

The changed string can be the following, if links exist

| 1 | 2 | 3 | 4 | 5 | 6 | 7 | 8 | 9 | 10 | | | |

FIGURE 11.5
Pictorial view of change of angle.

To deliver a new solution set, the combination of stones containing PS number of nodes is created randomly. The number of nodes that are expelled from either side of point hitting are denoted by FT. These are the following four combinations of operation:

- Case 1: Get In & No Removal; No solutions are removed and new solutions are added.
- Case 2: No Get In & Removal; Some solutions are removed from previous and no new solutions are added.
- Case 3: Get In & Removal; Previous solutions are removed and new solutions are added.
- Case 4: No Get In & No Removal; Old solutions are kept and no new solutions are added.

For the refusal of operation on the solution, the last combination is used and it denotes no operation. Another criterion is the point of application of the step.

The requirements of the application are the sole decider of the degree of the combination of operation. PS = FT is used for TSP problem because there are constant number of cities and they should not be repeated and this is used for satisfying the constraint of TSP.

There is a need to develop a strategy for quickening the solution convergence procedure and another considerable criterion is the point of hitting. For the application of pebble tossing step, any point of hitting can be selected for the problem of TSP, but discontinuous portions are considered as the best methods for problems such as path planning. Any point that is important and mapped positions can be a hit point for a continuous domain problem.

11.4.2 Rolling with Twigs

The Egyptian vulture has another bewildering expertise, which is moving with twigs, for the purpose of movement the objects can be rolled or other actions can be performed such as finding the position of feeble points or other part can be observed, which is confronting the floor. Power is also required in moving of the items; there is a need of the strong hold on the stick using the beak and the best possible stick should be used. In order to make or execute an object, there is a need of finding the correct stick and this is the inherited expertise of the bird. For the purpose of laying eggs and making brilliant nests there is a need of finding sticks that have correct twist at an ideal place. Delicate twigs can be sewed using mouth and some of the creatures have also given proof for it. In case of multiobjective optimizations, there is a need for changing the places of the variables to change the importance, and this may improve the fitness value by introducing the new solutions and all this can be described as the action of rolling the solution set. In order to find an appropriate matching for the arbitrary event, it will take a long time to complete the event as there is lesser number of hit points and number of alternatives is very large. These sorts of operations are very useful in such cases. In subsequent sections, there is a discussion on the probabilistic approach for certain parameter estimation and this can be used for graph-based problems and this is shown in Figure 11.4 [2] that shows the impact of the *moving with twigs*. In the problem of TSP, partial string can be shifted and this partial string is generated randomly but here we show the entire string for shifting (probabilistically, it can be resolved that whether the potential outcome will be either left or right). The adoption technique should consider that the validity of solution string should remain intact and it should not hamper and it is determined by the application only.

To control the implementation of *rolling with twigs*, there is a need of two parameters, which will be helpful in the mathematical formulation of the event. In order to determine the degree of operation these are the following two criteria.

DS = Degree of Roll, where DS = 0. Number of rolls are denoted by DS. Direction of rolling is denoted by DR, where probabilistically we have:

$$DR = 0 \text{ for right Rolling or Shift}$$
$$= 1 \text{ for left Rolling or Shift}$$

Deterministically the equation can be framed in following manner and 0 and 1 are randomly generated.

$$DR = \text{Left Rolling/Shift for RightHalf > LeftHalf}$$
$$= \text{Right Rolling/Shift for RightHalf < LeftHalf}$$

Secondary fitness of the right half of the solution string is denoted by right half and similarly the fitness value of left half is denoted by the left half.

For problems such as path planning in which we do not want to hamper the partially determined solution in constrained environment and for problems such as TSP, in which the global arrangement can be achieved effectively and the search methodology is more versatile because shifting the partial string folds the same information as the entire string. New solutions can be generated and the way exists between each node and separation is the Euclidean distance between them.

11.4.3 Change of Angle

To explore different avenues of tossing of stones and increment in the shot of breakage of the hard eggs, Egyptian vulture can perform an operation of change of angle and this increases the chances of breaking the eggs. Mutation step is represented as change of angle and in order to achieve a complete node sequence, unconnected link nodes can be turned for the desire of being connected. Such a stage is exhibited in Figure 11.5 [2]. If the fitness is improved after this step then only we will consider it.

This change of angle step depends on the number of nodes held by the string and it is a multipoint step in which the points are chosen by the local search and number of nodes should be considered. When there is a string that holds excessively numerous nodes then this step is a good option for local search and making a full path out of it because the pebble tossing step cannot be performed.

11.4.4 Brief Description of the Fitness Function

For the purpose of optimization and with regards to decision-making the fitness function is extremely important; however, it is seen that multiobjective optimization is the core of most of the graph-based problems, as all the steps are probabilistic, so the situation can be more worse and the number of iterations to reach a path are very large, so it is very difficult to delineate the better incomplete outcome from the others. Here, the secondary fitness function is discussed in brief. These are the following types of secondary fitness value:

1. Linked sequential nodes are numbered with numbers that represent the number of nodes of a portion that are connected together. The summation of the fitness or the number of nodes is used to ascertain the secondary fitness at that point. If more quantities of nodes will be connected together as a unit then it signifies large values of secondary fitness than other solution string. The secondary fitness will be of no use for TSP as it will be constant because there are connections between each node.

2. The amount of partial solution is the secondary fitness assessment for partial solutions and it is the connected portion that is available in that string and in comparison to disconnected one the probability of forming the complete path is higher in connected solutions. The secondary fitness value is the count of the connected segments introduced in the solution. Minimum will be given as best result and it is opposite to the previous strategy.

11.4.5 Adaptiveness of the Egyptian Vulture Optimization Algorithm

There are various adaptive experimentations, which are applied on problems having multi-variable in initial positions and it is reduced after the length of path and with the iterations. EVOA gives the alternative to such experimentation. Likewise, it can be adjusted for constant length combinatorial optimization and can be seen as a special instance of the algorithm. Versatile adaptivity of the operators is required in the circumstances when single or various amounts of nodes can be used to form the linkage between the two path fragments. Some great solutions on selected part of string can be produced using the operators such as *change of angle* and *rolling of twig* and can be combined to form a complete solution.

11.5 Applications of the Egyptian Vulture Optimization Algorithm

EVOA has numerous applications in the graph based problems, node-based continuity search, and optimization capacity issues. As there is no necessity of nearby search in the continuous space and its ability for various formations in the meantime forestalls loop development.

11.5.1 Results of Simulation of Egyptian Vulture Optimization Algorithm over Speech and Gait Set

EVOA is implemented over the speech dataset of UCI machine-learning repository and gait dataset of physionet.org. The outcomes from the studies are discussed in Figures 11.6 through 11.11 as follows:

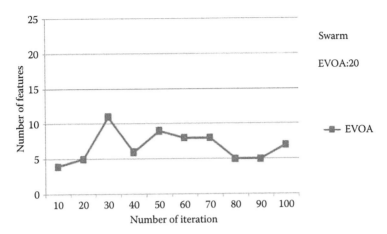

FIGURE 11.6
Plot of number of features versus number of iterations.

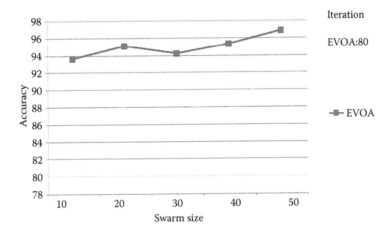

FIGURE 11.7
Plot of accuracy versus swarm size.

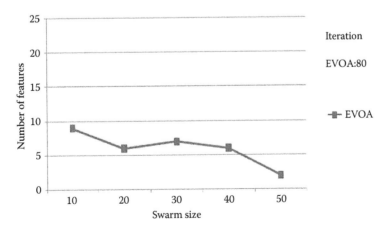

FIGURE 11.8
Plot of features versus swarm size.

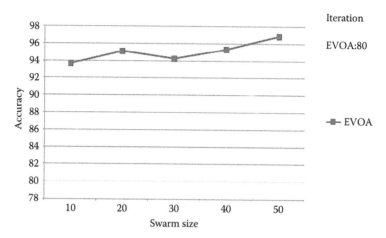

FIGURE 11.9
Plot of accuracy versus swarm size.

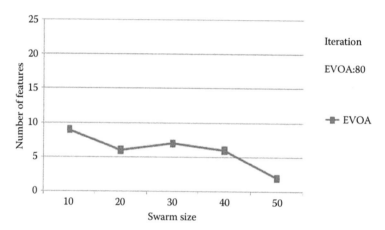

FIGURE 11.10
Plot of number of features versus swarm size.

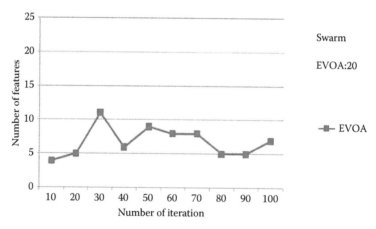

FIGURE 11.11
Plot of number of features versus number of iterations.

Exercises

1. What is the life style of an Egyptian vulture? In addition, how it can be used to form an optimization algorithm?
2. What are the steps followed by an Egyptian vulture while eating the eggs of other birds?
3. Draw the flowchart of EVOA and describe each step in brief.
4. What is the significance of pebble size and force of tossing in pebble tossing step?
5. What is the constraint of pebble size and force of tossing for solving TSP and why this constraint should be followed?
6. In the rolling of twigs step, what is the degree of rolling and direction of rolling?
7. Write the pseudocode of EVOA.
8. Describe secondary fitness function briefly.
9. Comment on the performance of EVOA over TSPLIB dataset.
10. Comment on the performance of EVOA over speech dataset of UCI repository.

Solved Questions

1. In any successful implementation of a nature-inspired algorithm there should be right balance between exploration and exploitation. How this factor is incorporated in EVOA?

Solution 1

In order to achieve the exploration, all three steps of the algorithm (pebble tossing, rolling with twigs, and change of angle) are performed. In pebble tossing step, new solutions are added and it is in the form of pebble, so this creates randomness in solution space. The rolling with twigs is also focused on the exploration as the solution string is rotated according to the degree of rolling and direction of rolling. In addition, the change of angle focuses on exploration as the solution set is reversed. So, all these three steps focus on exploration.

However, to incorporate exploitation, there is a provision of replacing top X% worst solution after each iteration; in addition, if the solution obtained all the three steps of EVOA is not better than its parent solution, then that change is not accepted. So, these two steps are focused on exploitation.

2. How EVOA can be used to solve TSP? Explain with the help of an example.

Solution 2

The following are the points of interest for applying EVOA utilized for TSP application.

> *Step 1*: Introduce N (taken as 20) solution strings with random generation of all the nodes present for a dataset without repetition. The string represents an arrangement of parameters, which overall represents a single state of allowable

arrangement. Here, the strings have the unrepeated nodes. In any case, generation of the solution strings included the tossing of pebbles step in which the sequence of cities is created through the arbitrary arrival of nodes and duplicate avoidance step. So, now the strings with no node will ascend to its greatest length and is equivalent to the dimension of dataset. Later, the tossing of pebbles venture in Step 3 will make changes in the string through a similar strategy; however, the length of string will stay in place taking care of the criteria legitimate TSP solutions.

Step 2: Initializing the primary fitness matrix (array), and secondary fitness is unnecessary as there are connected nodes. Calculate the initial fitness of the strings.

Step 3: Perform tossing of pebbles operation at the chosen or arbitrary point relying on the usage of deterministic approach or likelihood. If the chosen node(s) are optimized concerning the distance and is set with the node with which it has the least distance by performing out a local search in a limited zone (as continually seeking through the entire string is unrealistic, the search space is confined to some area between the position S_{max} and S_{min} where S_{max} < (measurement of TSP) and S_{min} > 0 and $S_{max} - S_{min}$ < threshold, and the threshold decides what number of greatest nodes are to be scanned and is held for the computational complexity of the algorithm. Placement and substitution of nodes is remunerated by shifting the nodes or the vacancies, whichever is desired. Acknowledge the new solution if that outperforms the old solution in fitness. This step will help in step by step diminishing the separation between the two nodes.

Step 4: Perform rolling of twigs operation on the chosen partition only, as the operation on the overall string would not be useful so far as the solution quality is concerned. It should likewise be possible by picking a specific length with node positions as L_{start} and L_{end} where L_{start} < L_{end} and $L_{end} - L_{start}$ < (Dimension of TSP) as though the entire string is shifted (right or left), the fitness estimation of the string stays unaltered; however, just the beginning and the last cities get changed.

Step 5: Perform change of angle operation through particular reversal of solution subset. This is some kind of additional exertion that is presented by the vulture for proficient outcome. It is the mutation operator for combination. Acknowledge the new arrangement if it outperforms the old solution in fitness. Same method of L_{start} and L_{end} is taken after.

Step 6: Fitness value of each string is evaluated and it will be the minimum distance connecting them.

Step 7: In the event that a new solution (inferred out of combination of operation[s]) is better, then replace the old, else do not.

Step 8: Best result is selected and it is compared with the global best. Global best is updated if the best result is more fit than it.

Step 9: With random initialization, X% of the worst solutions are replaced after every iteration (X depends on N and according to the requirement of exploration).

Step 10: Terminate if the number of steps are completed, else continue from Step 3.

NOTE: It is to be specified here that *tossing of pebbles* operation happens under the impact of some swarm of vulture bird agents each producing distinctive components at various positions and this quickens the solution generation much rapidly. The quantity of fowls in a swarm relies on the dimension of the dataset and how much rapidly the combination is required to be changed. However, the other two operations (*rolling of twigs* and *change of angle*) are likewise performed under impact of a swarm, yet the quantity of agents in swarms is relatively lower.

Computational Results of Simulation of Egyptian Vulture Optimization Algorithm over Traveling Salesman Problem Dataset

In this segment, we have given the outcomes comprising the mean, standard deviation (SD), best, worst, and mean error in a table while it is contrasted and the optimum value in dimensions of the datasets are denoted by dim. The simulation of the EVOA on TSP TSPlib datasets [20] extending from 16 to 280 dimensions gives that the range of 4.7% and 28.7% error when all the datasets are kept running for 25,000 iterations.

Figures 11.12 through 11.14 present the graphical perspective of the table with reference to the dimensions of the datasets as given in Table 11.1 [2]. The outcomes plainly present the capacity of the EVOA as a fruitful algorithm for TSP and experimentation has additionally shown that with the expansion in the quantity of dimension, the iteration number must increase to show the signs of an improved result.

EVOA is fit for combining the choices, yet it as usual relies on the likelihood. It demonstrated well on the TSP and the outcomes are converging toward the best with the quantity of iteration expanding with the expanded number of dimensions. The EVOA can be utilized for a wide range of node-based hunt problems and the fitness assessment methodology and validation checking procedure contrasts for every situation. Local searches such as setting the node with the closest node without arbitrary placement can, however, be a great methodology when confronting constraints less problems and can help in fast convergence, yet the continuous addition of such neighborhood searches may decimate the already set closest one.

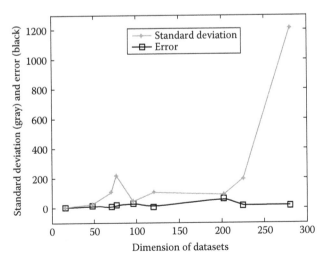

FIGURE 11.12
Plot for standard deviation and error.

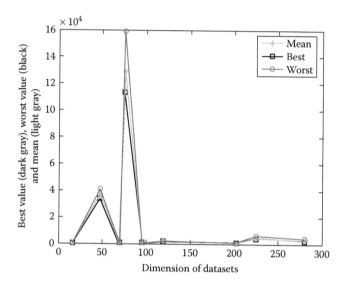

FIGURE 11.13
Plot best value, worst value, and mean.

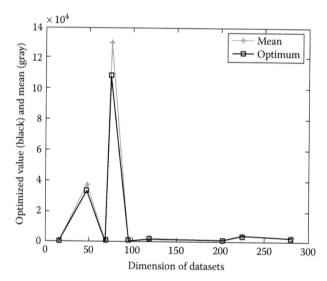

FIGURE 11.14
Plot for mean and optimum value.

TABLE 11.1

Table for Result

Datasets			EVOA				
Name	Dim	Optimum	Mean	SD	Best	Worst	Error
Ulysses16.tsp	16	74.11	77.56	1.16	75.16	79.53	4.6552
att48.tsp	48	3.3524e + 004	3.7119e + 004	23.89	3.3781e + 004	4.1136e + 004	10.7237
st70.tsp	70	678.5975	730.676	105.6	694	802	7.6744
pr76.tsp	76	1.0816e + 005	1.2941e + 005	219.1	1.1295e + 005	1.5898e + 005	19.6468
gr96.tsp	96	512.3094	659.11	46.7	599.2	1002	28.6547
gr120.tsp	120	1.6665e + 003	1.8223e + 003	106	1.7552e + 003	2.1825e + 003	9.3489
gr202.tsp	202	549.9981	886.92	92.2	650.2	1202	61.2587
tsp225.tsp	225	3919	4597	196	4216	6112	17.3003
a280.tsp	280	2.5868e + 003	2.9943e + 003	1213	2.7976e + 003	3.8236e + 003	15.7531

References

1. C. Blum and A. Roli, Metaheuristics in combinatorial optimization: Overview and conceptual comparison, *ACM Computing Survey*, 2003, 35, 268–308.
2. C. Sur, S. Sharma and A. Shukla, Solving travelling salesman problem using Egyptian vulture optimization algorithm—A new approach, *Language Processing and Intelligent Information Systems*, 2013, pp. 254–267.
3. J. Kennedy and R. Eberhart, Particle swarm optimization, *IEEE International Conference on Neural Networks*, 1995, 4, 1942–1948.
4. D. Karaboga, An idea based on honey bee swarm for numerical optimization. Technical Report TR06, Erciyes University, 2005.
5. A. H. Kashan, League championship algorithm: A new algorithm for numerical function optimization, *International Conference of Soft Computing and Pattern Recognition*, IEEE Computer Society, Washington, DC, 2009, pp. 43–48.
6. X. S. Yang and S. Deb, Cuckoo search via Levy flights, *World Congress on Nature & Biologically Inspired Computing*, IEEE Publication, 2009, pp. 210–214.
7. X.-S. Yang, A new metaheuristic bat-inspired algorithm. In: J. R. Gonz´alez, D. A. Pelta, C. Cruz, G. Terrazas, N. Krasnogor (Eds.) *NICSO 2010, SCI*, vol. 284, pp. 65–74. Springer, Heidelberg, Germany, 2010.
8. S. Kirkpatrick, C. D. Gelatt Jr. and M. P. Vecchi, Optimization by simulated annealing, *Science*, 1983, 220(4598), 671–680.
9. R. Storn and K. Price, Differential evolution—A simple and efficient heuristic for global optimization over continuous spaces, *Journal of Global Optimization*, 1997, 11(4), 341–359.
10. Z. W. Geem, J. H. Kim and G. V. Loganathan, A new heuristic optimization algorithm: Harmony search, *Simulation*, 2001, 76, 60–68.
11. K. Krishnanand and D. Ghose, Glowworm swarm optimization for simultaneous capture of multiple local optima of multimodal functions, *Swarm Intelligence*, 2009, 3(2), 87–124.
12. O. B. Haddad, A. Afshar and M. A. Mariño, Honey-bees mating optimization (HBMO) algorithm: A new heuristic approach for water resources optimization, *Water Resources Management*, 2006, 20(5), 661–680.
13. K. Tamura and K. Yasuda, Primary study of spiral dynamics inspired optimization, *IEEE Transactions on Electrical and Electronic Engineering*, 2011, 6, S98–S100.
14. S. H. Hamed, The intelligent water drops algorithm: A nature-inspired swarm-based optimization algorithm, *International Journal of Bio-Inspired Computation*, 2009, 1, 71–79.

15. M. H. Tayarani-N, M. R. Akbarzadeh-T, Magnetic optimization algorithms a new synthesis, *IEEE Congress on Evolutionary Computation*, 2008, pp. 2659–2664.
16. C. W. Reynolds, Flocks, herds and schools: A distributed behavioral model, *Computer Graphics*, 1987, 21, 25–34.
17. A. Kaveh and S. Talatahari, A novel heuristic optimization method: Charged system search, *ActaMechanica*, 2010, 213, 267–289.
18. A. H. Gandomi and A. H. Alavi, Krill herd algorithm: A new bio-inspired optimization algorithm, *Communications in Nonlinear Science and Numerical Simulation*, 2012, 17(12), 4831–4845.
19. Y. C. Liang and R. Josue, Virus optimization algorithm for curve fitting problems, *IIE Asian Conference*, 201106, 2011.
20. Dataset Library, http://elib.zib.de/pub/mp-testdata/tsp/tsplib/tsplib.html. http://www.math.uwaterloo.ca/tsp/inlinks.html.

12

Biogeography-Based Optimization

12.1 Introduction

There are numerous optimization algorithms, for example, genetic algorithms (GAs), ant colony optimization (ACO) algorithm, and so on, as discussed in Chapters 2 through 11. These algorithms are immensely used in fields related to image processing, bioinformatics, and so on, which can be used to solve problems having high complexity. Biogeography-based optimization (BBO) is a technique, which was formulated in 2008 by Dan Simon. The above-stated technique is based on the principles of biogeography. Biogeography is the study of the geographical distribution of biological organisms [1]. This algorithm has two concepts: migration and mutation.

Various application of biogeography to engineering, which has been popularized in the recent years are GAs, fuzzy logic, particle swarm optimization (PSO), neural networks, and other evolutionary algorithms. BBO relies on the principle of island biogeography, which encompasses biogeography that investigates the reasons affecting the richness of any species of isolated natural communities. Any area of suitable habitat, which is surrounded by an expanse of unsuitable habitat, is called an *island*. The richness of species of islands is affected by two main factors: migration and mutation. Migration includes two main operations: immigration and emigration. These processes encompass various factors such as size of the island, habitat suitability index (HSI), the distance to the nearest neighbor of an island, and so on. Vegetation, rainfall, climate, and so on are the various components, which are involved in HSI and affecting the survival of these species in a habitat. High emigration and low immigration rates are the properties of habitat with a high HSI value that are saturated with large number of species and vice versa. The same scheme is employed in BBO for carrying out the process of migration. At initial stage, we randomly generate a number of possible candidate solutions for the proposed problem and associated with each solution there will be HSI representing a habitat, which is a stock of suitability index variables (SIVs). These variables show the suitability of the habitat to which it belongs. A high HSI habitat is equivalent to a good solution and vice versa. Through migration, high HSI solutions share a lot of features with poor solutions [2].

Biogeography is a way of nature to divide species, quite similar to general problem solutions. For example, if we have a problem along with their candidate solutions, and the problem belongs to any area of life (medicine, engineering, urban planning, sports, economics, business, etc.), as long as we can measure the quantity of the suitability of a given solution. A good solution is similar to an island with a high HSI, whereas a low HSI represents an island with a poor solution. High HSI solutions oppose deviation at a higher rate than low HSI solutions and have tendency to share their features with low HSI solutions. (This means that the feature is alive both in high HSI solution and low HSI solution. This is similar to representatives of a species migrating to a habitat, while other representatives remain in their original habitat.) New features of the good solution are inculcated by the poor solutions. The quality of low HSI solution is raised by the addition of new features. This approach to solving the problem is called BBO.

Biology-based algorithms have certain features analogous with algorithms such as PSO and GAs. GA solutions *die* at the last phase of each generation as compared to the solutions of PSO and BBO, which live till the last (modifying their features as the optimization process progresses). PSO solutions have a built-in tendency to clump in homogeneous groups, whereas GA and BBO solutions do not form clusters.

12.2 Biogeography

Figure 12.1 illustrates a model of the abundance of species in a single habitat [3]. The immigration and the emigration rates are functions of the number of species in the habitat. Consider the immigration curve. The rate of immigration is at maximum when there are zero species in the habitat. As the number of species goes on increasing, the habitat becomes more and more crowded. Only few species are able to survive immigration to the habitat; thus, the rate of immigration decreases gradually. Thus, when immigration rate decreases to zero, the habitat is able to support largest number of species [4].

Considering the emigration curve, the emigration rate is zero if there are no species in the habitat. As the number of species in the habitat goes on increasing, it becomes more and more crowded, that is, more and more species leave the habitat in order to investigate other possibilities thus increasing the emigration rate. The emigration rate E is at maximum when the habitat contains the largest number of species that it can support.

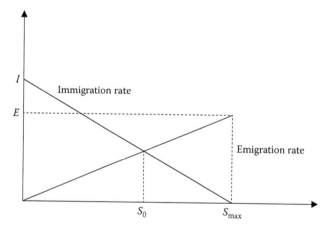

FIGURE 12.1
Species model of a single habitat. (Based on MacArthur, R. and Wilson, E., *The Theory of Biogeography*, Princeton University Press, Princeton, NJ, 1967.)

S_0, is the point at which the immigration and emigration rates are equal and is called the equilibrium condition. The temporal effects, however, may lead to excursions from S_0. For example, a sudden spurt of immigration may lead to positive excursions (caused, perhaps, by an unusually large piece of flotsam arriving from a neighboring habitat), or a sudden burst of speciation (such as a miniature Cambrian explosion). The introduction of an especially ravenous predator, disease, or some other natural catastrophe may cause negative excursions from S_0. It takes a long period for any species counts in nature to reach the equilibrium after a major perturbation [5,6].

Figure 12.1 shows the immigration and emigration curves as straight lines but, in general, they might be more complicated curves. Nevertheless, this simple model gives us a general description of the process of immigration and emigration. The details are changed as required.

Considering the probability P_s such that the habitat contains exactly S species, P_s changes from time t to time $(t + \Delta t)$ as follows:

$$P_s(t + \Delta t) = P_s(t)(1 - \lambda_s \Delta t - \mu_s \Delta t) + P_{s-1}\lambda_{s-1}\Delta t + P_{s+1}\mu_{s+1}\Delta t \tag{12.1}$$

where λ_s and μ_s are the immigration and emigration rates, respectively, having S species in the habitat. The conditions that must be true for the equation to hold having S species at time $(t + \Delta t)$ are

1. There were S species at time t, and no immigration or emigration occurred between t and $(t + \Delta t)$.
2. There were $(S - 1)$ species at time t and one species immigrated.
3. There were $(S + 1)$ species at time t and one species emigrated.

We assume that Δt is small enough such that the probability of more than one immigration or emigration can be ignored. Taking the limit of (1) as $\Delta t \to 0$ gives Equation 12.2 shown at the bottom of the page. We define $n = S_{max}$, and $P = [P_0 \ldots P_n]$, for notational simplicity. Now, we can arrange the equations (for $S = 0, \ldots n$) into the single matrix equation

$$P = AP \tag{12.2}$$

where the matrix A is given as Equation 12.5. For the straight line curves shown in Figure 12.1, we have

$$\mu_k = \frac{Ek}{n}$$

$$\lambda_k = I\left(1 - \frac{k}{n}\right) \tag{12.3}$$

Now, considering the special case, $E = I$.
In this case, we have

$$\Lambda_k + \mu_k = E \tag{12.4}$$

And the matrix A becomes

$$A = E \begin{bmatrix} -1 & \dfrac{1}{n} & 0 & \cdots & 0 \\ \dfrac{n}{n} & -1 & \dfrac{2}{n} & \ddots & \vdots \\ \vdots & \ddots & \ddots & \ddots & \vdots \\ \vdots & \ddots & \dfrac{2}{n} & -1 & \dfrac{n}{n} \\ 0 & \cdots & 0 & \dfrac{1}{n} & -1 \end{bmatrix} = EA' \qquad (12.5)$$

where A is defined by the above-mentioned equation.

12.3 Biogeography Based Optimization

Now, we will be discussing how the theory of biogeography mentioned in Section 12.2 can be successfully applied to the proposed optimization problems having a discrete domain.

12.3.1 Migration

Given a problem and a population of possible candidate solutions in which each solution is characterized as vectors of integers. Each integer present in the proposed solution vector is referred as a SIV. Further, let us consider that we are having different ways for determining the quality of the solutions. The habitat with high HSI value contains good solutions sustaining many species, whereas solutions having low HSI are poor and sustain only few species. HSI resembles *fitness* in other optimization algorithms which are based on population (such as GAs). Assuming that every solution (habitat) has a similar species curve (for simplicity, we are taking $E = I$), the S value presented by the given solution mainly depends on its HSI. S_1 in Figure 12.2 presents a solution having a low HSI value sustaining only few species as compared to S_2 representing a solution sustaining many species and having a high HSI value. The rate of immigration λ_1 will be higher for S_1 than the rate of immigration λ_2 for S_2. The rate of emigration for S_1, that is, μ_1 will be lower than the rate of emigration μ_2 for S_2.

In order to probabilistically share information between habitats, the rates of emigration and immigration, respectively, of each solution are used. Each solution is changed accordingly depending on other solutions with probability P_{mxl}. The rate of immigration λ of the selected solution is used to probabilistically determine if one should modify each SIV in that solution. If the solution S_i with a given SIV is selected for modification, then we use other's solutions emigration rates μ to probabilistically determine which of the solutions should migrate around the randomly selected SIV to solution S_i.

The global recombination of the breeder approach shares similar strategy with the BBO migration such as GA [7] and evolutionary strategies [8] in which a single offspring is contributed by many parents, as described in Chapter 3. The strategy of evolutionary algorithm uses global recombination, which is a reproductive process in order to develop new solutions, whereas BBO migration being an adaptive process changes the existing solutions. Incorporating some sort of elitism with population-based optimization algorithms helps in preserving the best solutions and also protecting them from being corrupted by immigration.

Algorithm 1   Pseudocode of Biography-Based Optimization Migration Operator

---

for $i = 1$ to N **do**
    for SIV $= 1$ to D **do**
        Generate a random value $r_1 \in [0,1]$
        If $\lambda_i > r_1$ **then**
            $H_i(\text{SIV})$ is selected
        else
            $H_i(\text{SIV})$ is not selected
        end if
        if $H_i(\text{SIV})$ is selected **then**
            Generate a random value r_2 and set Total_Sum $= r_2 \sum_{1}^{N} \mu_i$
            Set Temp_Sum $= 0$ and $j = 0$
            while Temp_Sum $\leq$ Total_Sum **do**
            $j = j + 1$
            Temp_Sum $=$ Temp_Sum $+ \mu_j$
            end while
            $H_i(\text{SIV}) = H_j(\text{SIV})$
        else
            $H_i(\text{SIV}) = H_i(\text{SIV})$
        end if
    end for
end for

12.3.2 Mutation

The HSI of a habitat is vigorously changed by the cataclysmic events and affects the count of any species to deviate from its equilibrium value (unusually large flotsam arriving from a neighboring habitat, disease, natural catastrophes, etc.). Thus, any random events can change a habitat's HSI. This is modeled in BBO as SIV mutation, and mutation rate is decided with the use of species count probabilities, which is governed by the given Equation 12.2. The equilibrium point on the species curve of Figure 12.2 shows that the counts of both low species and high species have relatively low probabilities. Being near to the balanced state the count of medium species has high probabilities. For example, considering the case where $S_{max} = 10$, the steady-state solution of Equation 12.2 is independent of the initial condition $P(0)$ and can be computed either numerically as shown in Equation 12.6. By summing all the elements of $P(\infty)$ we will obtain a sum of one and there might be a rounding error, with respect to its midpoint when $P(\infty)$ will be plotted it would be an even function.

$$P(\infty) \approx [0.001 \ \ 0.001 \ \ 0.044 \ \ 0.117 \ \ 0.205 \ \ 0.246 \ \ 0.205 \ \ 0.117 \ \ 0.044 \ \ 0.001 \ \ 0.001]^T \quad (12.6)$$

Each member of the population has an associated probability, a probability is indicated by it and the solution of the problem is expected to be apriori. Very low and very high HSI solutions are unpredicted encouraging larger number of medium HSI solutions. It is a matter of surprise that some solution is likely to be mutated by a given solution S, which existed and has a low probability P_s. When the solution has a high probability the chances of mutation to another one is reduced considerably. Mutation rate m can be used to implement it and solution probability has an inverse proportionality with it.

$$m(S) = m_{\max}\left(\frac{1-P_s}{P_{\max}}\right) \tag{12.7}$$

Here, we have a user-defined parameter, $m_{\max}$. The diversity among the population can be increased using this mutation scheme. There will be more domination of highly probable solution if we do not apply this modification scheme. Using this approach, low HSI solutions are more likely to be mutated and a chance of improvement is given to them. There are chances of less mutation for high HSI solutions. A high-elitism approach is used for saving the features of the habitat and in BBO it will be used for the solution, so the best solution is saved and even if its HSI is ruined, and if there is a need it can be reverted back. On both good and poor solution, the high-risk mutation process is used. Average solutions will be avoided for mutation and they are avoided because they are hopefully improving (except for the most probable solution, yet there is still some mutation probability). There is a problem-dependent mechanism for implementing mutation, and it is same as GAs. A randomly chosen sensor is replaced by a new randomly generated sensor and this is the process of mutation for sensor selection problem and it is discussed in Section 12.4. Alternate mutation schemes are not explored in this chapter, and all those mutation algorithms can be applied to BBO that have been already applied for GAs as described in Chapter 3.

12.4 Biogeography-Based Optimization Algorithm

Following algorithm gives the informal description of BBO algorithm:

Algorithm 2   Pesudocodes of Biogeography-Based Optimization Mutation

---

for $i = 1$ to N **do**
    Use λ_i and μ_i to compute the probability P_i and mutation rate m_i
    for SIV $= 1$ to D **do**
        Generate a random value R
        if $R < m_i$ **then**
            Generate a feasible value F in searching space
            H_i(SIV) $= F$
        end if
    end for
end for

Parameters for BBO should be initialized. A method is derived so that SIVs and habitats are mapped to problem solution, and this depends on the problem. Maximum species count S_{max}, m_{max} the maximum mutation rate and E and I are the maximum migration rates, and an elitism parameter (Section 12.3.1 last paragraph) is initialized. There are relative quantities such as maximum species count and the maximum migration rates. The behavior of BBO will remain the same if there is the same percentage of change in all these. For each solution, there will be the same relative amount of change in λ, μ, and species count if I, E, S_{max}, and m_{max} change.

Random set of habitats should be initialized; potential solution of the problem is corresponded by each habitat.

For each habitat, HSI is mapped to the λ immigration rate, S number of species, and the μ emigration rate.

Each nonelite habitat is modified probabilistically by using immigration and emigration as suggested in Section 12.3.1, each HSI is then recomputed.

For each habitat, there will be an update in probability of its species count. Then, each nonelite habitat is mutated using its probability and it is suggested in Section 12.3.2, and each HSI is recomputed.

In the next iteration go to step (3). After a predefined number of generations, this loop can be terminated; an acceptable problem solution can be founded later.

There is a need of verifying the problem solution after each modification of the habitat (steps 2, 4, and 5). If a feasible solution is not represented by it, then it should be mapped to the set of feasible solutions and there is a need of implementation of some method for it. The flow chart of BBO is shown in Figure 12.2.

12.5 Differences between Biogeography-Based Optimization and Other Population-Based Optimization Algorithm

A few of the distinct properties of BBO are presented in this section. Generation of *children* or reproduction is not involved in it but it should be noted that BBO is an optimization algorithm that is population-based. Reproductive strategies such as evolutionary strategies and GAs are clearly distinguished from it. As a new set of solutions are generated with each iteration in ACO but ACO is also clearly differentiated from BBO. On the other hand, in BBO, a set of solution is maintained from one iteration to the next; those solutions are adopted by relying on migration. PSO and DE have the strategies that are most common with BBO. From one iteration to the next solutions are maintained in those approaches, but neighbors of a solution are used in the learning and as the algorithm progresses it adapts itself. In PSO algorithm, each solution is represented as a point in space and for each solution a velocity vector is used to represent the change over time. There is no direct change in PSO solutions; there is a change in their velocity, and as an indirect consequence there is change in position (solution). Solutions are directly changed in DE, but particular DE solutions are changed based on other DE solutions and differences between them. There is a biological motivation in DE. There is a contrast between BBO and PSO, DE because migration from other solutions (islands) is used for direct changes in BBO solutions. Directly attributes (SIVs) are shared between BBO solutions and other solutions. The strength of BBO is proven using these differences between other population-based optimization methods and BBO. There are some open research questions: How the performance of BBO is differentiated from other population-based optimization methods because of the above differences? Due to these differences, which types of problems are most relevant for BBO? Initial explorations into BBO are presented in this chapter but questions for later work are left (Figure 12.2).

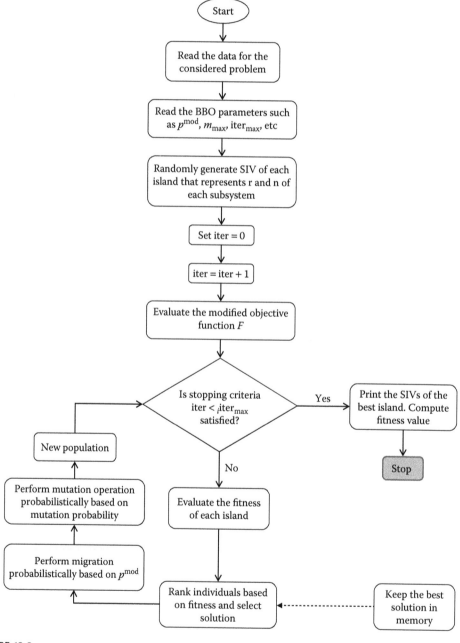

FIGURE 12.2
Flowchart of BBO algorithms.

12.6 Pseudocode of the Biogeography-Based Optimization Algorithm

01	Solution for each habitat
02	Solution Feature for each SIV
03	With probability $\propto \lambda_i$, H_i habitat should be selected
04	If $Ud(0 <comma> 1) < \lambda_i$ then
05	With probability $\propto \mu_j$, H_j is selected
06	If $Ud(0 <comma> 1) < \mu_j$ then
07	$H_i(SIV) \longleftarrow H_j(SIV)$
08	End if
09	End if
10	End for
11	End for

12.6.1 Some Modified Biogeography-Based Optimization Approaches

Some of the modified BBO approaches are proposed by researchers so that various drawbacks of the BBO can be eliminated and these are as follows:

12.6.1.1 Blended Biogeography-Based Optimization

In constrained optimization, a modified approach of BBO can be used and that is known as blended biogeography. In this algorithm, there is a proposal of a blended migration operator [5]. One SIV is not entirely replaced by another SIV but the features of both variables are combined in the blended migration operator. Thus, the degradation of the solution is prevented by this. The following equation describes the blended migration operation:

$$H_i(s) = \alpha H_i(S) + (1 - \alpha)H_j(s) \tag{12.8}$$

α is a real number and that ranges from 0 to 1. Emigrating and immigrating habitats are represented by H_j and H_i, respectively. If the value of α is closer to 1 then it is because of the more fitness of solution H_i than solution H_j and if the value of α is closer to 0 then it is because of the lesser fitness of solution H_i than solution H_j. Using several benchmark functions that are constrained, the performance of BBO, SGA, and PSO is compared with the blended BBO. In comparison to others, the blended BBO performed very well. It was also an observation that the performance of the algorithm is significantly impacted by the value of α. The problem that is to be solved should decide the value of α. On various benchmarks the best performance is given when 0.5 was used as the value of α.

Following shows the blended BBO algorithm.

1. Mutation probability and migration probability should be defined.
2. Population should be initialized.
3. For each candidate in the population, the emigration rate and immigration rate should be calculated.
4. Based on the immigration rate, island is to be selected for modification.

5. On the emigration rate, Roulette wheel selection should be used, an island should be selected from which the SIV is to be emigrated.

6. Choose a random SIV among all the emigrating islands selected and then execute migration mechanism using blended migration operation defined by Equation 12.4.

7. Carry out the mutation operation according to the mutation probability calculated for each island.

8. Define the fitness value for each of the islands in the habitat. Terminate if the fitness criteria are satisfied, else return to Step 3.

12.6.1.2 Biogeography-Based Optimization with Techniques Borrowed from Evolutionary Strategies

Evolutionary strategies (ES) are derived from the natural evolution process of the nature; they implement the selection, mutation, and recombination operators resulting in the formation of new individuals in the new generation. The selection process allows the users to evaluate other members of the whole population in which the users choose the best members among the whole population. BBO algorithm implements this technique in order to increase the probability of finding the best island. In the new hybrid BBO/ES algorithm, migration procedure is kept similar to the basic BBO algorithm, whereas the mutation procedure is applied on the population defined according to the mutation probability for the child islands. After that the fitness of the whole population members is calculated. The best individuals based on the fitness are passed to the next generation. The new BBO/ES algorithm is observed to show better performance than the original BBO for evaluation of several benchmark functions.

12.6.1.3 Biogeography-Based Optimization with Immigration Refusal

In the BBO algorithm, a solution with high fitness may also take features from another solution having low fitness. However, it may result in the degradation of the fitness of the immigrating island. Biogeography-based optimization/immigration refusal (BBO/RE) algorithm overcomes this drawback of the original BBO algorithm. Thus, in the new algorithm, an immigrating island takes features from islands that have fitness greater than some threshold and also than that of the immigrating island. The two improved algorithms, that is, BBO/ES and BBO/RE, can be merged together to form (BBO/ES/RE). The new algorithm is then tested on benchmark functions and is observed to significantly outperform BBO algorithm as well as BBO/RE algorithm with a poor result obtained in only one iteration out of many. BBO/ES/RE also showed better results than BBO/ES in two out of fourteen testing iteration. It can be inferred from the results that ES techniques and immigration refusal show a strong effect on the BBO algorithm.

12.6.1.4 Differential Evolution Combined with Biogeography-Based Optimization

Differential evolution (DE) is a simple optimization algorithm for global optimization. It is a population-based direct search algorithm and real-valued parameters. DE algorithm scored a second runner-up position in the first international contest on evolutionary computation for solving real-valued functions among the test suite of optimization functions. DE algorithm efficiently explores the search space in order to locate the global minima. However, when it comes to exploiting solutions, the DE algorithm is observed to give slow results. Summary of various variants of BBO is represented by Table 12.1.

TABLE 12.1

Key Features of Various BBO Approaches

Algorithm	Features
BBO	• An efficacious algorithm for optimization • Does not take additional computational time • Proficient in exploiting solution • Solutions does not die at the end of each generation like other optimization algorithm
Blended BBO	• Avert the degradation of the solutions • Significantly classy than BBO • Excellent for solving constrained optimization problem
BBO/ES	• Make use of evolutionary strategy idea to select the finest islands after a mutation • Reduced computational cost as the fitness of the parents is already computed in previous generation • Considerably better than BBO and BBO/RE
BBO/RE	• Does not show appropriate performance over BBO • Tuning the parameters the immigration refusal is an area where there is scope for future work
BBO/RE/ES	• Almost 10 times better than BBO • Since immigration refusal idea does not have appropriate impact on BBO, BBO/ES is more commendable than BBO/ES/RE
BBO/DE	• Combine the advantage of DE and BBO • Superb at exploring the search space and exploring the outcomes • Significantly better than BBO/DE

12.7 Applications of Biogeography-Based Optimization

12.7.1 Biogeography-Based Optimization for the Traveling Salesman Problem

The TSP problem [9] is easier to define but very difficult to solve. The problem was to find a shortest possible path traversing a given list of cities with the given cost of a path between the two cities. The path generated had to be such that each city was visited only once and the path returned to the origin node. For n number of cities in the map, a number of n! paths could be generated between the source and destination. As the number of cities becomes large, the computational complexity in terms of number of possible solutions also increases with an exponential amount, thus creating difficulty in even today's high-speed computer to generate the path. The sequence of SIVs in the BBO algorithm is used when solving TSP rather than a single SIV. These sequences of SIVs determine the fitness of the population and the best sequence of the SIV defines the solution of the problem. In a new approach, a modified BBO algorithm is implemented by implementing parallel computation [10]. The new approach gives 24% better output for the problem solved using a GA. The modified algorithm is explained as follows:

1. Distribute the population into n shares. The shares formed are to be distinct and are then sent to various subislands.
2. Determine the best SIV combination and assign it as the new sequence for the original island.

3. According to the performance for each of the subisland, define the emigration and immigration population such as the basic BBO.

4. If the expected criteria are met, terminate the algorithm, otherwise move to Step 2.

The TSPBBO algorithm is as follows:

1: Initialize m_{max}, generation number N, generate the initial path habitats H^n randomly
2: Evaluate the HSI for each path individual in H^n
3: Sort the population from best to worst based on cost
4: Initialize the generation counter $N = 1$
5: While the halting criterion is not satisfied do
6: For each path individual, map the HSI to the number of species
7: Calculate the immigration rate λ_i and the emigration rate μ_i for each path $H(i)$
8: Calculate P_s
9: If rand $< \lambda_i$ and $\mu_i < $ sum(μ_i)
     If habitat $H(i)$ selected
       Crossover $(H(i), H(i+1))$
     End if
   End for
10: Calculate mutation rate m_i
11: Mutate each habitat in H^n with mutation rate m_i
12: Evaluate H^n, update HSI
13: Sort H^n based on cost
14: Keep the first two best individuals
15: $N = N + 1$
16: End while

12.7.2 Biogeography-Based Optimization for the Flexible Job Scheduling Problem

BBO algorithm implements the concept of migration among animals in order to solve complex optimization problems. The flexible Job Shop scheduling problem is solved by searching a solution through the migration operators of BBO and an optimum/near-optimum solution for the problem is obtained.

12.7.3 Biogeography-Based Optimization of Neuro-Fuzzy System; Parameters for the Diagnosis of Cardiac Disease

Cardiomyopathy is considered as a medical term for a heart disease in which the heart muscles become thick, rigid, and enlarged. Due to this, the myocardial cells tend to become electrically unstable, leaving the heart vulnerable to arrhythmias or failure. There are usually two forms of cardiomyopathy, dilated and hypertrophic, meaning the enlargement or swelling of atria. A new approach to diagnosis of the disease implements a neuro-fuzzy network implementing BBO algorithm. The P wave features of the disease are identified using the neuro-fuzzy networks that are helpful for the diagnosis of the disease.

In addition, for improving the training of the approach the algorithm implements a concept of opposition-based learning in addition to the BBO algorithm. The clinical data of ECG are then trained using the BBO algorithm. The observed results of the training define the new modified algorithm as an effective method for the diagnosis of cardiomyopathy.

12.7.4 Biogeography-Based Optimization Technique for Block-Based Motion Estimation in Video Coding

BBO is a recently developed bio-inspired optimization technique and is a population-based evolutionary algorithm. The main idea behind this algorithm is the solution sharing features based on the probability defined by the solutions' fitness values. The algorithm thus shows good exploitation property. In a particular instance of literature, the BBO algorithm is tested for the block video estimation in the field of video coding. A new technique for motion compensation in video coding is developed and it predicts the current video frame by analyzing the previous frame (reference frame). This technique uses the temporal redundancy among the successive frames. The new technique is then compared to the existing solutions in the literature and the observations are analyzed using experimental techniques. The results obtained make the new proposed solution a competitive one among the most optimized solutions in the current literature with reduced complexity.

12.7.5 A Simplified Biogeography-Based Optimization Using a Ring Topology

This is another approach, which implements BBO algorithm to obtain a simplified method for optimization. The original BBO algorithm uses the concept of global topology that allows migration between any two pair of habitats. The solutions (habitats) are optimized through local ring topology, wherein every habitat gets connected to two habitats in the solution space. The migration operator can then be applied to the neighboring habitats. This strategy is easily implemented and affects significantly toward the improvement of the solutions by enhancing the search capability and takes out the solutions stuck in the local optima. Several benchmark problems are tested on this approach and the algorithm is observed to show effective results for the problem.

12.7.6 Satellite Image Classification

BBO algorithm is known to efficiently identify features from various types of images obtained from the satellite. These images can be of water, vegetation, rocky areas, urban, and so on. The BBO algorithm considers the multispectral band in the image as a single SIV and the standard deviation observed in the pixels as HSI.

12.7.7 Feature Selection

Another area of significance for the BBO algorithm is the feature selection area. For various applications, feature selection is applied as a preprocessing measure, so as to reduce the data's dimensionality. For face recognition applications, the BBO algorithm is often applied in order to reduce the extracted image data's dimensions. Feature selection is also implemented on DNA microarray data in order to identify cancer cells. The problem tackled by the BBO algorithm is that of overfitting of data due to large number of genes than the number of samples in the microarray dataset. This problem is also denoted as the curse of dimensionality and this problem is solved through the use of BBO algorithm.

12.8 Convergence of Biogeography-Based Optimization for Binary Problems

BBO studies the concept of biogeography in which the species migrate between habitats to develop an evolutionary strategy for optimization problems. An earlier work in the area derived a finite Markov chain model of the BBO algorithm implemented on binary problems. The results of the application were recorded and their analysis revealed that BBO algorithm when implemented using only mutation and migration showed no signs of convergence on global optimum. However, the BBO algorithm implemented with elitism, converged to the global optimum as the newer population was defined from the best individuals from the previous population. These results were quite advantageous and can be applied to obtain better results than artificial neural networks (ANNs). The literature states several differences between GA and the BBO algorithms; however, it has been observed in the studies that the BBO and the GA are known to show similar convergence properties.

Solved Questions

1. Discuss an application of BBO algorithm in medical field.

    Here we present the task of diagnosing Alzheimer's disease with the help of BBO instead of using the conventional back propagation algorithm of ANN.

    The dataset used is cross-sectional dataset from OASIS that provides brain imaging data that are freely available for distribution and data analysis.

    Overall features for the algorithm are derived out of patient's history of psychological and clinical tests. Records log neurotransmitters amount in blood, physical movement patterns, dementia state, change in size of temporal lobes, acalculia, alexia, and other characterizing features that turn out to be determiners in this case. All these attributes are fed in as the input to the algorithms, which are first trained according to the hybrid algorithm.

    Following parameters are used as input to the proposed algorithm (Table 12.2).

    The following diagrams highlight the result obtained as a result of experiments. (Figures 12.3 and 12.4).

TABLE 12.2

Main Features in Dataset

Age	Age at the time of acquisition (years)
Sex	Sex (male or female)
Education	Years of education and socioeconomic status assessed by Hollingshead index of social position and classified into categories from 1 (highest status) to 5 (lowest status)
MMSE score	Ranges from 0 (worst) to 30 (best)
CDR scale	0 = no dementia, 0.5 = very mild AD, 1 = mild AD, and 2 = moderate AD
Atlas scaling factor	Computed scaling after (unitless) that transform native-space brain and skull to the atlas target (i.e., the determinant of the transform matrix)
eTIV	Estimated total intracranial volume, cm^3
nWVB	Expressed as the percent of all voxels in the atlas-masked image that is labeled as white or gray matter by the automated tissue segmentation process

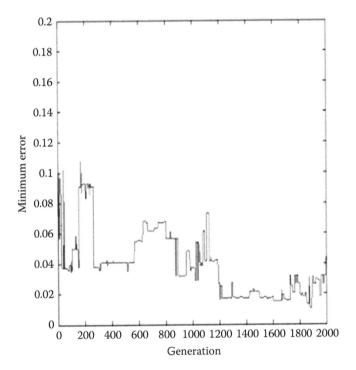

FIGURE 12.3
Performance of BGOA in training ANN for small dataset (OASIS dataset).

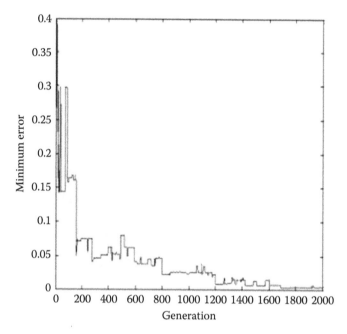

FIGURE 12.4
Performance of BGOA in training ANN for large dataset (OASIS dataset).

2. Apply the blended migration operator given that immigrating and emigrating parameters are 12 and 13.87, respectively. Take $\alpha = 0.76$.

Using,

$$H_i(s) = \alpha H_i(S) + (1-\alpha)H_j(s)$$

We have

$$H_i(s) = (0.76 * 12) + (1 - 0.76) * 13.87$$

$$H_i(s) = 9.12 + 3.322$$

$$H_i(s) = 12.422$$

Unsolved Questions

1. Comment on the usefulness of the BBO algorithm.
2. Highlight the major limitations of BBO algorithm clearly citing the factors that contribute to these limitations.
3. Highlight the basic principles of biogeography as applicable in the real world.
4. In what ways is diversity achieved in BBO algorithm?

5. How is the balance between exploration and exploitation achieved in BBO algorithm?
6. What is the physical significance of SIV in the BBO algorithm?
7. How does blended BBO improve on the performance of standard BBO? Explain.
8. How is differential evolution going to help BBO in giving better results?
9. What factors do contribute to the effective convergence of BBO algorithm?
10. How is immigration rate λ, and emigration rate μ manifested in the BBO algorithm?

References

1. A. Wallace, *The Geographical Distribution of Animals* (Vol. 2), Adamant Media Corporation, Boston, MA, 2005.
2. T. Wesche, G. Goertler, and W. Hubert, Modified habitat suitability index model for brown trout in southeastern Wyoming, *North American Journal Fisheries Management*, 1987, 7, 232–237.
3. R. MacArthur and E. Wilson, *The Theory of Biogeography*, Princeton University Press, Princeton, NJ, 1967.
4. C. Darwin, *The Origin of Species*, Gramercy, New York, 1995.
5. I. Hanski and M. Gilpin, *Metapopulation Biology*, Academic, New York, 1997.
6. A. Hastings and K. Higgins, Persistence of transients in spatially structured models, *Science*, 1994, 263, 1133–1136.
7. H. Muhlenbein and D. Schlierkamp-Voosen, Predictive models for the breeder genetic algorithm: I. Continuous parameter optimization, *Evolutionary Computation*, 1993, 1, 25–49.
8. T. Back, *Evolutionary Algorithms in Theory and Practice*, Oxford University Press, Oxford, UK, 1996.
9. P. Pongcharoen, W. Chainate, and P. Thapatsuwan, Exploration of genetic parameters and operators through travelling salesman problem, *ScienceAsia*, 2007, 33(2), 215–222.
10. D. du, Biogeography-based optimization: Synergies with evolutionary strategies, immigration refusal, and kalman filters, ME Thesis, Cleaveland University, Department of Electrical and Computer Engineering, 2009.

13

Invasive Weed Optimization

Invasive weed optimization (IWO) algorithm is yet another swarm insight-based algorithm, which has some trademark making it unmistakable and one of a kind from other conventional social conduct-based calculations. While particle swarm optimization (PSO) and bacteria rummaging rely on the social conduct-based impact of other cospecialists, IWO algorithm operators do not show such an impact-based character. Rather they have their extraordinary nearby hunt qualities in view of standard deviation for the best particles, which are required to be available at the ideal and close ideal spots. Another critical trademark is the proliferation, which helps in a new era and in particular irregular introduction for new specialists, scattered at different purposes of the workspace. An end step adjusts the consolidated impact of spatial scattering and generation elements. Next, we will talk about the established IWO algorithm and afterward the discrete variation of IWO algorithm called the discrete invasive weed optimization (DIWO) algorithm.

13.1 Invasive Weed Optimization

IWO algorithm was initially presented by Mehrabian and Lucas in 2006 [1]. IWO algorithm is propelled from the advance and development of weed plants all through the field in a way, which supports their development and are furnished with sufficient requirement that is required for smooth working and improvement. The most intriguing part of the algorithm is the regular sense of the calculation for neighborhood that seeks alongside the correct usage of multiplication variable for nearby quests. The neighborhood hunt is included with haphazardness and gives satisfactory extension investigation; however, with restricted open door chosen by the specific components. IWO algorithm has the accompanying interesting operations for drifting over the hunt space:

Population initialization: Initially, the data structures for the arrangement set of the issue are defined and initialized. In the event, they are introduced by maintaining the limitations containing domain/range alongside other essential data representations. Aside from that this operation puts the invasive weed specialists at different areas of the scan space through arbitrary era for the variable estimations.

Reproduction: In this progression, the mother weed (speaking of the weed operators of the past emphases) produces posterity through the procedure of irregular era inside a stipulated space computed by the spatial scattering stage. However, the quantity of delivered offspring for a mother weed depends on the fitness of the mother weed in a linear fashion. The quantity of created offspring depends or is somewhat run between a base and the most extreme, and the quantity of operators changes straightly, considering that the best-fitted specialist will deliver the greatest, whereas the most exceedingly awful-fitted operator will create the base. The rest will be obliged some place in the middle of the most extreme and the base. This results in more offspring for a better-fitted weed and the other way around.

Spatial Dispersal: Now, the creation of offspring must proceed with their arrangements at different areas of the workspace. A question emerges in the matter of how these areas will be chosen? This is done through a scientific definition appeared beneath as σ_{iter}. The variety σ_{iter} is the territory around the specialist, which goes about as the locale where the weed can embed their offspring and the developing procedure of the weed must proceed. In this progression the mother weed scrambles the seeds for offspring all through the search space; though randomly yet restricted to this standard deviation condition.

$$\sigma_{iter} = \frac{(iter_{max} - iter)^n}{(iter_{max})^n}(\sigma_{initial} - \sigma_{final}) + \sigma_{final} \tag{13.1}$$

where $\sigma_{initial}$ and σ_{final} are the initial and final standard deviation values considering normal distribution, $iter_{max}$ is the maximum possible iteration for the program to run, and σ_{iter} is the standard deviation for the current iteration denoted by iter, whereas n is the nonlinear modulation index. Here, the main motive is to generate agents as seeds are being randomly scattered with a distribution as normal over the search space.

Competitive Exclusion: In this progression, weeds are sorted concerning the wellness and a specific rate of the slightest fitted ones are barred predominantly relying on the quantity of new ones made amid the proliferation criteria for wellness of every mother weed. This instrument utilizes the standard of *survival of the fittest* offering opportunity to every single other plant with lower wellness to replicate, and if their offspring have great wellness, they can get by in their posterity's presence.

13.1.1 Invasive Weed Optimization Algorithm in General

Instate M operators each with M Memory or M factors for One Solution
Begin $i = 1$ to Iteration_countFor Loop
Begin $k = 1$ to Iteration_End
Begin $j = 1$ to N for Loop [N is the quantity of mother weed agents]
Select Solution Subset for Agent j where Solution Subset $\in \{1, 2,... ...., M\}$
Ascertain the Spatial Parameters in view of the wellness parameters
Perform Spatial Dispersal
Perform Reproduction to produce new weed from mother weed
Perform Competitive Exclusion to Compensate for the Excess weed operators
End j for Loop
Assess Fitness for every Weed
Upgrade Global Best if Better Weed Found

Supplant the less fitted ones with the more fitted ones.
End k for Loop
End i for Loop

IWO algorithm chiefly suits ceaseless area issues; for example, multidimensional numerical optimization of benchmark conditions, optimization of scientific likeness models, and so on. Henceforth, in the accompanying area a discrete adaptation of the IWO algorithm is presented named as DIWO algorithm, which supports the discrete occasion-based optimization issues such as combinatorial optimization issues and diagram-based issues [2].

13.1.2 Modified Invasive Weed Optimization Algorithm

Another changed rendition of the IWO algorithm was presented by supplanting the spatial scattering condition to a more versatile one. The variety of that changed IWO algorithm is with the accompanying condition

$$\sigma_{iter} = \frac{(iter_{max} - iter)^n}{(iter_{max})^n} * |\cos(iter)| * (\sigma_{initial} - \sigma_{final}) + \sigma_{final} \tag{13.2}$$

The modified equation has some extra term, which can bring about some more variations in the spread of the offspring weed. This will enhance the local search and better optimized search can be reached. However, the condition for its success can depend highly on its factors that is denoted by n and cos(iter). The $|\cos(iter)|$ introduce a degree of variation in σ, which aids in exploring the better results fast and keeps the new solutions to spread out of the search space in which σ may be generally large. As search demands unexpected moves and the revolving or rather cyclic characteristics of cos(iter) have made the modified algorithm much better with better convergence.

13.2 Variants of Invasive Weed Optimization

Another variation of the IWO algorithm has been examined underneath. The principal adjustment is the attributes of spatial dispersal with irregular choice of arrangements from a neighboring hypercube in the discrete space of arrangements around the plant with a normal distribution as opposed to utilizing the standard deviation for scattering. Utilizations of the discrete space for appropriation of posterity, however, have not made the calculation allowed to use in discrete applications [3] such as occasion-based calculation. Be that as it may, this changed rendition has utilized some new approach with the spatial dispersal stage. However, there are difficulties in characterizing the neighboring hypercube and the determination of a suitable spatial dispersal for posterity seeds that ought to lie close to their parent plant.

The next position for dispersion can be either $v_i - 1$ or $v_i + 1$ and likewise with the increasing number of dimensions D we can have more number of options. Cells are then selected and changed randomly to make a new seed using the following equation:

$$L(\text{iter}) = \text{floor}\left(\frac{N_c - 1}{\ln(\text{iter}_{\max})}\right) X \ln(\text{iter}) + 1 \qquad (13.3)$$

Here, we go for decreasing the standard deviation for a weed when the target work estimation of a specific weed nears the minimum objective function value of the present population so that the weed scatters its seeds inside a little neighborhood of the suspected optima.

13.2.1 Modified Invasive Weed Optimization Algorithm with Normal Distribution for Spatial Dispersion

---

Initialize M agents with each M Memory or M variables for One Solution
Start $i = 1$ to Iteration_count for Loop
Start $k = 1$ to Iteration_End
Start $j = 1$ to N for Loop [N is the number of mother weed agents]
    Select Solution Subset for Agent j where Solution Subset $\in \{1, 2, \ldots \ldots, M\}$
    Compute the number of seeds corresponding to its fitness
    Randomly select the seeds from the feasible solutions around the parent plant in
    a neighborhood with normal distribution
    Perform Reproduction to generate new weed from mother weed
    Perform Competitive Exclusion to Compensate for the Excess weed agents
End j for Loop
Evaluate Fitness for each Weed
Update Global Best if Better Weed Found
Replace the less fitted ones with the more fitted ones.
End k for Loop
End i for Loop

Discrete Invasive Weed Optimization: DIWO algorithm is the discrete adaptation of IWO and has a similar number of operations yet the sequence and the implication of the operations have been adjusted. This is done as such that it can be used for the path planning in diagram situations and other discrete generation of events and the operations can work with occasions (in such issues the vital element is that the arrangement of the events holds a huge significance). The following divide fundamentally manages the operations of DIWO algorithm and depictions of those operations concerning particular illustrations, which are talked about beneath.

Population Initialization: The introduction bit is the principal operation and unless determined, there might be some emphasis in which this operation might be required over and over for better open door and reinstatement. However, this progression for the most part manages the ground arrangement for the principle operations to execute and get worked on and generally is same as that of the conventional one, yet the information representation for issues such as way-arranging issues (here represented as mother weed agents, which is quite similar to chromosomes that are represented in GA) is versatile and increments with cycles (as here some substantial succession of events/nodes are chosen) until it achieves the goal with a valid pattern of connected nodes in between or the iteration has finished. They are very unique in relation to other advanced course of action, sort of issues, and diverse for other sort of discrete issues such as diagram-based issues and combinatorial issues in which there are some different sorts of prerequisites and have settled dimensional representation of arrangements (Figure 13.1) [4].

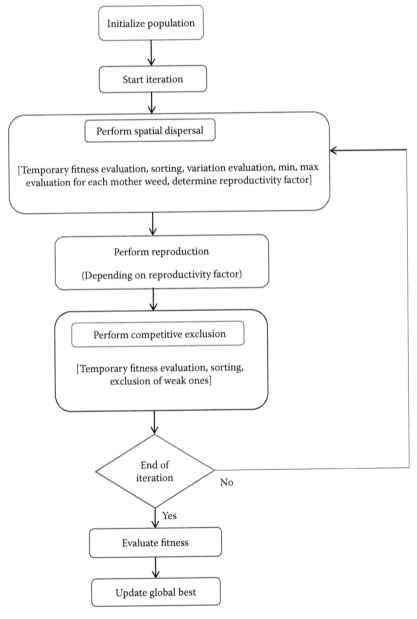

FIGURE 13.1
Flowchart for discrete invasive weed optimization algorithm.

Spatial Dispersal: Spatial dispersal operation is one sort of estimation and location of the forward or next inquiry movement in the network based on graphs. Furthermore, evaluation of the temporary fitness for each data chromosome is carried out using $\sum$(edge parameters/number of nodes). Then, sorting of the mother weeds is done according to the temporary fitness evaluated with the best fitted representing the first element. Now, the value for MIN is selected (fixed from the starting of iterations and in the simulation it is considered as an integer such as 2 or 3, but it also depends on the average number of edges connected to a node). Though the MAX is also limited and is reinitiated to MAX/2, every time the MAX reaches or crosses the maximum, so MAX can achieve a maximum value, but MAX is determined using the equation MAX = (MIN + σ_{iter}) for each iteration. But if MAX crosses the maximum then MAX takes the value as MAX = MAX/2 to honor the maximum bound of MAX and

$$MAX = (MIN + \sigma_{iter}) \qquad (13.4)$$

where σ_{iter} is given in the equation as

$$\sigma_{iter} = \left[\frac{|iter_{max} - iter|^n}{(iter_{max})^n}(\sigma_{final}) + \sigma_{initial} \right] \qquad (13.5)$$

and forethought must be given for the value of $\sigma_{initial}$ and σ_{final} such that the MAX value stays within boundaries.

This function is accounted as an integer value. If n is not a factor of 2 then the difference of $iter_{max}$ and iter is kept inside the mod to take into consideration that the $\sigma_{initial}$ value increases. With the increment of iteration, the value of σ_{iter} slowly decreases. The reason for this is that as the end nodes are near, consequently there are lesser requirements for exploration.

Standard deviation is the range and the MAX value determines how many offspring (or child plant) a mother weed can have with the best-fitted weed given as MAX permission, worst fitted as MIN, and the others or intermediates with relevant proportions. In the next step, this value is required for the process of reproduction. This value is named reproductivity factor.

Reproduction: In this progression of multiplication, the mother weeds forward its posterity to the following nodes as though the weeds are spreading and the ways are chosen randomly from the accessible ways. Number of generations or the quantity of posterity relies on the reproductivity factor. Be that as it may if the reproductivity factor is more than the accessible ways, then weed posterity may take after a similar way and later may separate. Mind must be taken so that there must be sufficient information structure to handle the abundance created posterity. The posterity weeds turn into the mother weed for the following cycle and the posterity weed acquire the sought way of the mother weed.

Competitive Exclusion: In this progression, again all the weeds are assessed with temporary wellness and sorted and the least fitted ones are wiped out to keep in place the quantity of weeds for the cycles.

13.2.2 Discrete Invasive Weed Optimization in General

Initialize M agents with each k Memory (for maximum k nodes)
Place M Weed Agents at Source Event or at Random Event
Start $i = 1$ to Iteration_countFor Loop
Start $j = 1$ to N For Loop
Select Agents $\in$ {1, 2,......., M}
Perform Spatial Dispersal
Perform Reproduction
Perform Competitive Exclusion
End j For Loop
Evaluate Fitness for each if Solution is complete
Update Global Best
Replace the less fitted ones with the more fitted ones.
End i For Loop

13.3 Invasive Weed Optimization Algorithm for Continuous Application

The IWO algorithm has operations, which will oversee the scientific condition based and has the accompanying stream of calculation.

13.3.1 Invasive Weed Optimization for Mathematical Equations

Initialize M agents with each M Memory or M variables for One Solution
Start $i = 1$ to Iteration_countFor Loop
Start $k = 1$ to Iteration_End
Start $j = 1$ to N For Loop [N is the number of mother weed agents]
    Select Solution Subset for Agent j where Solution Subset $\in$ {1, 2,......., M}
    Calculate the Spatial Parameters based on the fitness parameters
    Perform Spatial Dispersal
    Perform Reproduction to generate new weed from mother weed
    Perform Competitive Exclusion to Compensate for the Excess weed agents
End j for Loop
Evaluate Fitness for each Weed
Update Global Best if Better Weed Found
Replace the less fitted ones with the more fitted ones.
End k for Loop
End i for Loop

13.3.2 Discrete Invasive Weed Optimization Algorithm for Discrete Applications

DIWO algorithm can be utilized for diagram-based issue and in this area, we have demonstrated the algorithmic stream for various sorts of uses.

13.3.2.1 Invasive Weed Optimization Algorithm Dynamics and Search

The flow of the IWO algorithm is represented for the most part by the spatial dispersal operation. The customary IWO algorithm was for the most part impacted by that progression and the pursuit procedure was spread with the standard deviation included. The dimensionality of the standard deviation is an additional favorable position than the conventional biopropelled calculation in which the right mix for high-dimensional arrangements is less inclined to happen much of the time and this is the reason IWO algorithm is so fruitful for optimization for consistent area look space. In conclusion, it must be said that the discrete form of the IWO algorithm is likewise administered by this spatial dispersal operation and this made it feasible for the IWO algorithm to be connected on chart-based calculation and the combinatorial optimization issues. Another part of the IWO algorithm is the capability in the nearby hunt and is free from the slope of any influencer; however, is yet powerful and conclusive and it is conceivable to play out a superior neighborhood seek when it is the region of an ideal pinnacle. In general, this calculation is profoundly viable and requires less computational power and it can be joined with another calculation to get the best out of the supporting calculations [5]. A case of the cross breed of the IWO algorithm and PSO is given underneath. This calculation depends on scientific optimization in light of constant factor-based arrangement of measurement *D*.

13.3.2.2 Hybrid of IWO and Particle Swarm Optimization for Mathematical Equations

---

Initialize *M* agents with each *M* Memory or *M* variables for One Solution [6]
Start $i = 1$ to Iteration_countFor Loop
Start $k = 1$ to Iteration_End
Start $j = 1$ to *N* for Loop [*N* is the number of mother weed agents]
      % Invasive Weed Optimization Algorithm
      Select Solution Subset for Agent *j* where Solution Subset $\in \{1, 2,\ldots\ldots, M\}$
      Calculate the Spatial Parameters based on the fitness parameters
      Perform Spatial Dispersal
      Perform Reproduction to generate new weed from mother weed
      Perform Competitive Exclusion to Compensate for the Excess weed agents
End *j* for Loop
Evaluate Fitness for each Agent
Update Global Best if Better Weed Found
Replace the less fitted ones with the more fitted ones.
Start $j = 1$ to *N* for Loop [*N* is the number of mother weed agents]
      % Particle Swarm Optimization
      Select Solution Subset for Agent *j* where Solution Subset $\in \{1, 2,\ldots\ldots, M\}$
      Calculate the inertia (ω), acceleration factor c_1, c_2 and $r_1, r_2 \in [0,1]$
      Calculate the Velocity for each variable *j* of Solution Subset
      Calculate the New Position of each variable *j* of Solution Subset
End *j* for Loop

Evaluate Fitness for each Agent
Update Global Best if Better Weed Found
Replace the less fitted ones with the more fitted ones.
End k for Loop
End i for Loop

13.4 Related Work

As of now, huge concentration has been given to the utilization of developmental calculations to take care of reception apparatus optimization issues [6]. Different designs have been picked and in each such setup, distinctive parameters have been taken to enhance the reception apparatus structures.

Frameworks of nonlinear conditions have emerged in many fields of tremendously useful significance. Some unmistakable fields incorporate designing, mechanics, solution, science, and apply autonomy. All things considered, settling this arrangement of straight conditions is especially required. The arrangement includes discovering each one of the arrangements of the polynomial conditions introduced in the framework.

Keeping in mind the end goal to explain this arrangement of conditions, a crossover system has been proposed. This strategy consolidates IWO and DE. The thought is as per the following:

1. Every beginning weed, contingent on its wellness esteem, delivers new seeds.
2. The created seeds are then haphazardly circulated everywhere throughout the hunt space.
3. Following the above-mentioned stride, just the weeds with better wellness survive and create seed, whereas others are killed.

Figures 13.2 through 13.4 [7] shows that the convergence speed of DEIWO is faster than that of IWO, and the optimization precision higher than that of IWO.

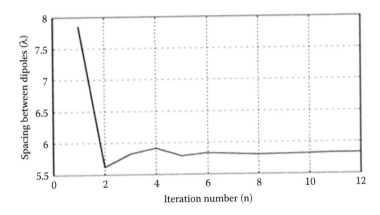

FIGURE 13.2
Dipole spacing in linear array versus IWO iteration number.

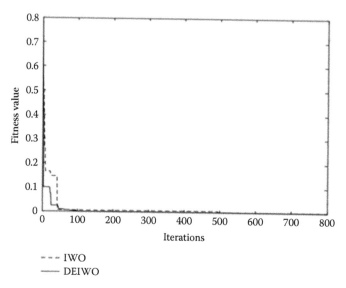

FIGURE 13.3
Results of 500 iterations.

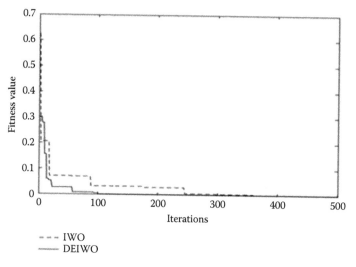

FIGURE 13.4
Results of 800 iterations.

A modified IWO calculation [8] has additionally been exhibited for optimization of multitarget flexible job shop scheduling problems (FJSSPs). The criteria here stays to minimize the maximum completion time (make span), the aggregate workload of machines, and the workload of the individual machine.

The problem of FJSSP has a place with the class of NP-hard family. It introduces two difficulties. The first is the watchful assignment of every operation to the machine, and the second one is the scheduling of this set of operations with a specific end goal to improve our measurements. In addition, the limitations while tackling this issue are:

1. Machines are independent of one another.
2. A machine can undergo a breakdown during its operation.
3. Jobs submitted to the machines are independent.
4. Each job j_i can start at the date $t = 0$ and the total number of operations to perform is greater than the number of machines.

The target here is to discover a schedule, which has a minimum make span, a minimum aggregate workload of machines, and a minimum workload of the basic machine. The total of these three measurements is taken to be the goal work. Taking after are the introduction parameters for the IWO calculation.

In [9], IWO and PSO-based power framework stabilizer (PSS) is intended to analyze their tuning exhibitions.

Eigen-value-based objective function is considered for the tuning of PSSs. This is done to improve framework damping of electromechanical mode. The execution of the proposed IWO-based PSS and PSO-based PSS is tested and exhibited under various unsettling influences for a four-machine case-control framework. The outcomes demonstrate that both IWO-based PSS and PSO-based PSS configuration can effectively damp out oscillations in this way enhancing the stability of the framework.

To choose the best stabilizer parameters, which will improve the power framework flow the most, an eigen value-based objective function is considered.

$$J = -\min\left(\frac{\text{real (eigen values)}}{\text{abs (eigen values)}}\right) \tag{13.6}$$

Now, the goal here is to maximize the minimum of the damping ratio for a certain parameter set. Maximizing the minimum damping ratio will help the cause of improving the system's overall damping. Therefore, the design problem can be illustrated as the following optimization problem:

$$\text{Maximize } J \tag{13.7}$$

Subject to the constraints,

$$K_i^{\min} \le K_i \le K_i^{\min} \tag{13.8}$$

$$T_{1i}^{\min} \le T_{1i} \le T_{1i}^{\max} \tag{13.9}$$

$$T_{2i}^{\min} \le T_{2i} \le T_{2i}^{\max} \tag{13.10}$$

$$T_{3i}^{\min} \le T_{3i} \le T_{3i}^{\max} \tag{13.11}$$

$$T_{4i}^{\min} \le T_{4i} \le T_{4i}^{\max} \tag{13.12}$$

where i is the number of stabilizers considered and J is the objective function defined in Equation 13.6 (Figures 13.5 through 13.7) [9].

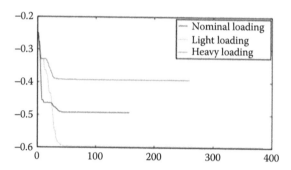

FIGURE 13.5
Experimental results under different loading conditions.

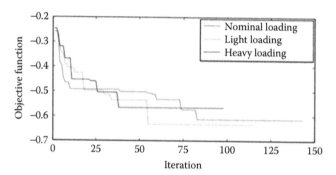

FIGURE 13.6
Variation of objective value with number of iterations.

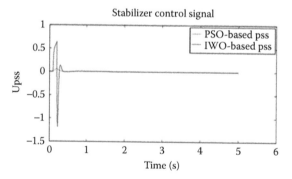

FIGURE 13.7
Variation of Upss value with respect to time for the two algorithms.

13.5 Summary

In this chapter, we have explained the IWO algorithm in most of the parts and some of its variations that are accessible in the nature are described. The peruser will get a reasonable view of how the execution has been upgraded for application. In addition, the discrete form of the IWO algorithm has been depicted and how it can be utilized for diagram-based applications. The striking elements of this chapter are as per the following:

- IWO algorithm and its changed variants are talked about.
- DIWO algorithm for discrete applications such as combinatorial optimization issues have been talked about.
- The adjusted rendition demonstrates how the calculations are changed for better hunt and execution.
- The presentation of the discrete form has opened up the calculation for a few different applications.

Solved Questions

1. Illustrate the utilization of IWO calculation to unravel the adaptable employment shop planning (FJSSP) issue.

   The objective of the solution is to find a schedule, which has a minimum make span, a minimum total workload of machines, and a minimum workload of the critical machine.

   These metrics are

   1. The make span: Cr1
   2. The total workload of machines: Cr2
   3. The workload of the most loaded machine: Cr3

   The sum of these three metrics is taken to be the objective function.

The calculation is actualized in C++ on an Intel(R) Core(TM) i3 CPU M370@2,40 GHz machine. The outcomes got by utilizing are likewise contrasted and the ones found in [10]. Taking after outcomes are gotten.

Unsolved Questions

1. Discuss how IWO calculation emerges from other transformative calculations.
2. Project the fundamental prerequisites to getting precise outcomes utilizing IWO calculation.
3. Discuss the characteristic constraints with IWO calculation. What steps could be taken to relieve them?
4. What elements decide the compelling meeting of the IWO calculation?
5. Discuss 'Invasive' in Invasive Weed Optimization.
6. What is the physical noteworthiness of nonstraight regulation record in IWO calculation?
7. In the setting to IWO calculation, the consolidated impact of spatial scattering and generation figures should be adjusted. Why?

8. In what way does IWO calculation support nearby hunt. Clarify.

9. Discuss the method of reasoning behind the spatial scattering.

10. Does the underlying populace influence the focused avoidance? Legitimize your reply.

References

1. A. R. Mehrabian and C. Lucas, A novel numerical optimization algorithm inspired from weed colonization, *Ecological Informatics*, 2006, 1, 355–366.

2. M. R. Ghalenoei, H. Hajimirsadeghi and C. Lucas. Discrete invasive optimization algorithm and its application to cooperative multiple task assignment of UAVs, *Proceedings of the 48th IEEE Conference on Decision and Control*, Shanghai, China, December 2009, pp. 1665–1670.

3. J. Kennedy and R. C. Eberhart, A discrete binary version of the particle swarm algorithm, *Systems, Man, and Cybernetics, 1997. Computational Cybernetics and Simulation, 1997 IEEE International Conference on*, vol. 5, pp. 4104–4108, October 12–15, 1997.

4. C. Sur and A. Shukla, Discrete invasive weed optimization algorithm for graph based combinatorial road network management problem. In *Proceedings of the 2013 International Symposium on Computational and Business Intelligence (ISCBI)*, August 2013, pp. 254–257.

5. M. Sahraei-Ardakani, M. Roshanaei, A. Rahimi-Kian and C. Lucas, A study of electricity market dynamics using invasive weed optimization, *Proceedings of IEEE Symposium on Computational Intelligence and Games*, Perth, Australia, December 2008, pp. 276–282.

6. H. Hajimirsadeghi and C. Lucas, A hybrid IWO/PSO algorithm for fast and global optimization, *Proceedings of EUROCON 2009*, St. Petersburg, Russia, May 2009, pp. 1964–1971.

7. Y. Zhou, Q. Luo and H. Chen, A novel differential evolution invasive weed optimization algorithm for solving nonlinear equations systems, *Journal of Applied Mathematics*, vol. 2013, 18, 2013. doi:10.1155/2013/757391.

8. S. Mekni and B. C. Fayech, A modified invasive weed optimization algorithm for multi-objective flexible job shop scheduling problems, *Computer Science & Information Technology*. pp. 51–60, 2014.

9. A. Ahmed and B. M. R. Amin, Performance comparison of invasive weed optimization and particle swarm optimization algorithm for the tuning of power system stabilizer in multi-machine power system, *International Journal of Computer Applications*, 2012, 41(16), 29–36.

10. W. Xia and Z. Wu, An effective hybrid optimization approach for multi-objective flexible job-shop scheduling problems, *Journal of Computers and Industrial Engineering*, 2005, 48, 409–425.

14

Glowworm Swarm Optimization

14.1 Introduction

Glowworm swarm optimization (GSO) algorithm is a swarm intelligence algorithm, which follows the behavior of lightning worms in order to solve complex mathematical problems. The algorithm was introduced by Krishnanand and Ghose [1]. GSO algorithm is very efficient in solving multimodal equations and has been used in many applications such as clustering, routing, swarm robotics, image processing, localization, and so on. GSO works on the mechanism of the physical behavior of insects called glowworms. The agents for the solution are seen to as glowworms, which contain luciferin, a light emitting compound found in many firefly species. The glowworms calculate the fitness of their current position in the solution space; get to use the objective function into a luciferin value, which they then broadcast to all their neighbors. The glowworm finds its neighbors and then calculates its position changes by manipulating its adaptive neighborhood, which is bordered by its sensor range. Every glowworm gets attracted by the brighter glow of their other neighboring glowworms and they then select, using a probabilistic method, a neighbor that has a luciferin value more than its own and moves in the direction of it. The higher the intensity of luciferin, the better is the location of glowworm in the search space. These movements—derived only from the local information and selective neighbor interaction enables the glowworm swarm to divide into disjoint subgroups that converge on multiple peaks of a given multimodal function. This algorithm is considered similar to that of other swarm intelligent solutions such as particle swarm optimization (PSO) and ant colony optimization (ACO) as discussed in the previous chapters but there are several notable differences in its working from the PSO:

- The velocity update equation of PSO includes a memory element, whereas in GSO no information is taken from the memory.
- In PSO, the particle movement directions are according to their own and global best positions in the previous iteration, whereas in GSO agent's directions are aligned along the line of sight among the neighbors.
- PSO achieves the dynamic neighborhood by evaluation of the first k neighbors. GSO uses the k neighbors only as implicit parameter to control the variable decision domain's range.
- Traditional PSO only performs numerical optimization, whereas traditional GSO can effectively detect multimodal solutions other than numerical optimization.

GSO employs a swarm of agents that are deployed randomly in the search space. These agents are named after glowworms, an insect larva which shows a glowing

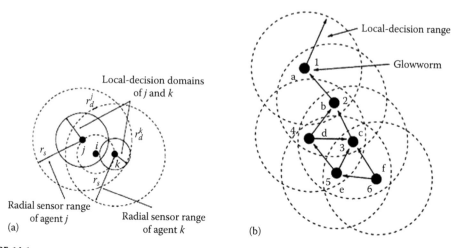

FIGURE 14.1
Variable neighborhood of glowworms. (a) Sensor range of different agents; (b) Relative luciferin level of each agents. (Taken from the original paper). (From Krishnanand, K. N. and Ghose, D., *Swarm Intell.*, 3, 2, 87–124, 2009.)

behavior through bioluminescence. They contain a luminescent quantity called as luciferin, which enables them to emit light and whose intensity is dependent on the associated luciferin contained in them. They interact with other glowworms present in their variable neighborhood. The neighborhood is a local decision domain with a variable neighborhood range r_{id} and is bounded by radial sensor range r_s ($0 < r_{id} \leq r_s$). For a glowworm i, glowworm j is considered as its neighbor, if j is under the neighborhood range of i and j has higher luciferin value to that of i. The decision domain allows for selective interactions between neighbors and helps in formations of disjoint subgroups. Every glowworm gets attracted by other glowworms having brighter glow within their neighborhood. The decision taken by a glowworm in GSO is only according to other glowworms in its neighborhood. For example, in Figure 14.1a [2], the agent i is within the sensor range of j as well as k. It can be seen that j and k have different neighborhood sizes, and only j uses i's information. In Figure 14.1b, a directed graph derived from the relative luciferin level exhibited by each swarm agent and on the availability of local information is shown. Based on a probabilistic mechanism, a glowworm selects a neighbor that contains higher luciferin level than its own and then moves toward it.

The movements, being based on local information and selective neighbor interactions, help the glowworm swarm to get partitioned into disjoint subgroups moving forward and meet at multiple optima of the multimodal functions.

14.1.1 The Algorithm Description

The description of GSO algorithm is presented in this section. The GSO algorithm works in several phases, which decide the movement of glowworms in the solution space resulting in optimal solutions for the problem. The GSO algorithm starts by adding a population of n glowworms in the search space randomly making sure they are well dispersed. Initially, all the glowworms are given an identical quantity of luciferin l_0. The iterations of the algorithm start with a luciferin-update phase and then followed by a movement phase according to a transition rule.

The phases in the GSO algorithm are explained as follows:

Luciferin-update phase: The luciferin-update phase changes the luciferin values of glowworms depending on the function value at the position of glowworm. During this phase, each glowworm appends to its original luciferin level, a luciferin quantity according to the fitness of its location in the objective function space. In addition, some part of the luciferin value gets removed from the glowworms to simulate the degradation of luciferin with time. The luciferin update rule can be defined as follows:

$$\ell_i(t+1) = (1-\rho)\ell_i(t) + \gamma J[x_i(t+1)] \tag{14.1}$$

Here, $l_i(t)$ is defined by the luciferin level of the glowworm i at time t, ρ can be given as the luciferin decay constant (value $0 < \rho < 1$), γ can be given as luciferin enhancement constant, and $J[x_i(t)]$ is the objective function's value of i's position at time t.

Movement phase: In this phase, glowworms decide to move toward other glowworms in their neighborhood, which have higher luciferin value than their own. In Figure 14.1b, a directed graph with six glowworms is shown, which decide their position after each iteration according to their corresponding luciferin levels and their availability of the local information. In the figure, it was shown that four glowworms (i.e., a, b, c, and d) show higher luciferin levels than glowworm e. Now, e is present in the neighborhood on only c and d and thus can move in only two possible directions. For a glowworm i, the probability of moving in the direction of a neighbor j is defined as

$$p_{ij}(t) = \frac{\ell_j(t) - \ell_j(t)}{\sum_{k \in N_i(t)} \ell_k(t) - \ell_i(t)} \tag{14.2}$$

where $j \in N_i(t)$, $N_i(t) = \{j : d_{ij}(t) < r_d^i(t); l_i(t) < l_j(t)\}$ contains an ordered set of the neighbors of glowworm i during time t, $d_{ij}(t)$ can be defined as the Euclidean distance between glowworms i and glowworm j during time t, and $r_d^i(t)$ can be defined as a variable neighborhood range of glowworm i during a time t. The probability of a glowworm i choosing a glowworm $j \in N_i(t)$ can be given as $p_{ij}(t)$ in Equation 14.2. After that, the movement of the glowworm can be defined using the following discrete-time equation:

$$x_i(t+1) = x_i(t) + s\left(\frac{x_j(t) - x_i(t)}{\|x_j(t) - x_i(t)\|}\right) \tag{14.3}$$

where $x_i(t) \in R^m$ denotes the position of glowworm i, at time t, inside the m-dimensional real space. Here, R^m denotes the Euclidean norm operator, and s (>0) as the step size.

Neighborhood range update property: An agent i is associated with a neighborhood whose radial range r_i^d is dynamic in nature ($0 < r_i^d \leq r_s$). A dynamic neighborhood is chosen because of some reasons. The glowworms many a times chose the local information only to decide their movements and that results in number of peaks being a function of radial sensor range. Thus, if the radial sensor range is covering the whole search space, the agents tend to move in a global optimum and prevent falling in a local optimum. It is usually the case where there is no information given about the objective function beforehand, so it is not possible to fix a chosen neighborhood value that works for different cases on the landscape of the objective function. In some cases, a chosen neighborhood range r_d might work better on objective functions in which minimum interpeak distance is less than the r_d, whereas it may not work well where it is larger. So, the GSO employs the use of an adaptive neighborhood range, so it can effectively detect multiple peaks in multimodal objective functions.

Let us assume that r_0 is the initial neighborhood range of a glowworm (i.e., $r_i^d(0) = r_0 \; \forall i$).

The rule for adaptively updating the neighborhood range for each glowworm can be given as follows:

$$r_d^i(t+1) = \min\left\{r_s, \max\left\{0, r_d^i(t) + \beta(n_t - |N_i(t)|)\right\}\right\} \tag{14.4}$$

where β is a constant parameter and n_t is a parameter used to control the number of neighbors.

The GSO discussed earlier can be summarized as the following:

Steps for Glowworm Swarm Optimization Algorithm

Procedure (GSO) (Problem Dimension, agents, Constraints, Search Space)

Initialize n agents with m dimensions as $\{x_1, x_2, \ldots, x_m\}$, x_i denoting position of ith glowworm.

Set step size as s.

Deploy these agents randomly in the solution space.

For $i = 1$ to n do:

Set $l_i(0) = l_0$ initializing initial luciferin value (time = 0) as l_0

End For

$r_i^d = r_0$.

Initialize maximum iteration number as it_max and $t = 1$

While ($t <$ it_max)

        For each glowworm i do: #luciferin Update phase

                Calculate luciferin value $l_i(t)$

        For each glowworm i do: #Movement Phase

                $N_i(t) = \{j : d_{ij}(t) < r_i^d(t); l_i(t) < l_j(t)\}$;

                for each glowworm $j \in N_i(t)$ do:

$$p_{ij}(t) = \frac{\ell_j(t) - \ell_j(t)}{\displaystyle\sum_{k \in N_i(t)} \ell_k(t) - \ell_i(t)} \tag{14.2}$$

$j =$ select_glowworm (p);

$$x_i(t+1) = x_i(t) + s\left(\frac{x_j(t) - x_i(t)}{\|x_j(t) - x_i(t)\|}\right) \tag{14.3}$$

        $r_i^d(t+1) = \min\{r_s, \max\{0, \text{rid}(t) + \beta(n_t - |N_i(t)|)\}\}$;

    End for

    Increment t by 1;

    End while

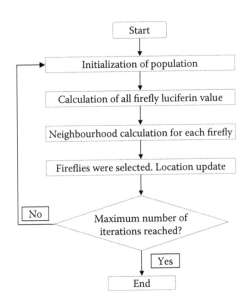

FIGURE 14.2
Flowchart of GSO algorithm.

A detailed flowchart of GSO algorithm is given in Figure 14.2.

14.2 Variants of Glowworm Swarm Optimization Algorithm

There have been several variants for the GSO algorithm in the recent years, which tried to tackle any shortcomings that the original GSO encountered or to make the algorithm work in a specialized problem statement. The various improved variants of the traditional GSO algorithm will be discussed in this section, and their features will be illustrated. It will be highlighted how the main variations and introduced operational features have been better for understanding the underlying dynamics of the particles of the algorithm and exactly how the performance of the algorithm has been for various applications.

14.2.1 Hybrid Coevolutionary Glowworm Swarm Optimization (HCGSO)

The GSO algorithm can be successfully used to solve the function optimization problems. The problem of solving a system of linear equations can be easily converted into a function optimization problem and the application of GSO can be tried. However, it is seen that the convergence speed and the local search ability of the algorithm are quite slow due to dynamic decision domain of the GSO algorithm. A simplex search method is known to show high local search ability on these types of problems. Hence, it can be seen that the hybrid of these two algorithms has the potential to remove the shortcomings of GSO. In this section, a new algorithm of GSO with simplex search method is explained in which the simplex search method is a local operator, which is then added to the GSO algorithm in order to speed up the local search ability of GSO [3].

14.2.1.1 Transformation of the Problem

Suppose that the system of nonlinear equations is as follows:

$$
\begin{aligned}
f_1(x_1, x_2, \cdots, x_m) &= 0, \\
f_2(x_1, x_2, \cdots, x_m) &= 0, \\
&\cdots \\
f_K(x_1, x_2, \cdots, x_m) &= 0.
\end{aligned}
\tag{14.5}
$$

Solving Equation 14.5 is equivalent to solve a function optimization problem as follows:

$$
\min P(X) = \Sigma f_i(x_1, x_2, \cdots, x_m)
\tag{14.6}
$$

Hence, solving Equation 14.6 is transformed into solving a set of values $x^* = (x_1, x_2, ..., x_m)$, which makes equation $P(x^*) = 0$.

14.2.1.2 The Process of HCGSO

Now our task is to minimize Equation 14.5. On account of the efficiency and accuracy of simplex search method that have a great relationship with the initial point, it is not having parallelism and only gets a solution one time, whereas GSO can overcome its disadvantages; however, the efficiency and accuracy of GSO are low and convergence speed slows. The simplex search method is embedded to GSO for creating a new hybrid algorithm. In the new algorithm, if a glowworm has neighbors, it moves to its one neighbor with a probability, if it has no neighbors, then it shows it is local optimum. Its quality is better and can be used as the initial point of a simplex search method. Then, the information of two algorithms is utilized in each other. We can define the fitness function as follows:

$$
J(x) = \frac{1}{[1 + P(x)]}
\tag{14.7}
$$

Its specific implementation steps are illustrated as follows:

Step 1: Set the initial luciferin ℓ_0, initial decision domains range r_0^d, circular sensor range r_s, neighbor number n_t, moving step of glowworm s, and iteration number of simplex search T. Set the stop criteria of HCGSOSSM; for example, max_iter as the maximum iteration.

Step 2: Randomly initialize a population and the population has n glowworms.

Step 3: Calculate fitness value and current location of each glowworm i, gain a corresponding luciferin value and its neighbors set $N_i(t)$.

Step 4: If $N_i(t)$ is not empty, calculate p_{ij} and $x_i(t+1)$; otherwise, $x_i(t)$ is used as the initial point of Cauchy iteration method and implement Cauchy iteration method.

Step 5: Update r_{id}.

Step 6: (Judge the stop criteria) If the stop criteria are satisfied, the algorithm ends and outputs the results; otherwise, go to Step 3.

This new algorithm has the following characteristics:

1. Structure complementarity, enriches the structures, and enhances parallelism. Simplex search method's structure is simple, it is the serial structure; GSO is the parallel structure. After mix, the structures are rich. Those enhance parallelism.
2. Behavior complementarity raises the convergence speed. Simplex search method iterates quickly, but it has no population information, easy to fall into the local minimum; the speed of GSO is slow, but it has certain overall situation ability. After the mix, simplex search method enhances the convergence speed and rate of GSO.
3. Weakens the stringency of parameters. The two algorithms have some parameters and the below experiments show that the dependence on the parameters of the new algorithm is not strong.
4. Weakens the influence of the initial points.
5. As GSO can solve multimodal optimization problems, for multisolutions of system of nonlinear equations, HCGSO can get all multisolutions at a time.

14.2.2 Glowworm Swarm Optimization with Random Disturbance Factor

In order to avoid GSO trapped in local optimal, give it the ability to expand the search scope, explore new areas. A random disturbance factor (R) at the movement updates stage is introduced, and the formula is as follows:

$$x_i(t+1) = x_i(t) + s\left(\frac{x_j(t) - x_i(t)}{\|x_j(t) - x_i(t)\|}\right) + \sigma * \text{randn} \tag{14.8}$$

where randn is a random number in $[-1,1]$, σ is the weighting factor of random disturbance.

14.2.3 Glowworm Swarm Optimization Algorithm Based on Hierarchical Multisubgroup

Glowworm swarm optimization algorithm based on hierarchical multisubgroup (HGSO) algorithm is a two-layer structure model. The whole search space is divided into multiple regions for the underlying layer; meanwhile, the swarm of glowworms is also split into many subgroups, and each subgroup is distributed to each region. To ensure that glowworms can disperse evenly on the entire space and can find the optimum in all regions, the chaotic map is used to initialize the position of the subgroups. At the same time, in order to ensure the exchange of information among different subgroups, the radial sensor range r_s is set to a large value. Then combine the peak-captures of each subgroup into a top swarm and capture multiple peaks of a given multimodal function. To improve the convergence performance of the top swarm, adaptive step size is presented that change with the number of iterations. In addition, to maintain the diversity of the population, selection and crossover are introduced. The underlying subgroup of the algorithm can search optimum of each region through GSO algorithm, and the top glowworms can quickly and accurately search multiple optimum of multimodal function.

14.2.3.1 Improved Logistic Map

Chaos has better ergodicity and randomness. In the GSO algorithm, the position of the glowworms is initialized by a chaotic map that cannot change the randomness of GSO algorithm but can improve the diversity and ergodicity of glowworms, so the improved. Mathematical expression of improved logistic map is given as [4]

$$y_{id}(t+1) = \begin{cases} 4\mu y_{id}(t)[0.5 - y_{id}(t)], & 0 \le y_{id}(t) < 0.5 \\ 1 - 4\mu[1 - y_{id}(t)][y_{id}(t) - 0.5], & 0.5 \le y_{id}(t) \le 1 \end{cases} \tag{14.9}$$

where $y_{id}(t) \in (0,1)$, $i = 1, 2, ..., i$, $d = 1, 2, ..., D$, $\mu = 4$. The improved logistic map is generated that has better randomness and glowworms can be more evenly distributed in the whole search space as compared to the general chaotic sequence.

14.2.3.2 Adaptive Step Size

K. N. Krishnanand proposed [4] that step size is an important parameter that affects the convergence of GSO algorithm and changes with the number of iterations. During the course of iteration, the distance between each glowworm and the brightest glowworm in its own neighbor set is changed. In the initial stage, we need a large step size to accelerate the search speed, but with the development of iteration, step size need reduce to prevent the glowworms from skipping the optimal solution. So, the fixed value of step size is not efficient to GSO algorithm and it must be adjusted with the number of iterations. A new step size update equation can be given as

$$s(t) = s_{max} - \frac{t(s_{max} - s_{min})}{t_{max}} \tag{14.10}$$

where s_{max} and s_{min} are, respectively, the step size of maximum and minimum, t_{max} indicates the maximum number of iterations.

14.2.3.3 Selection and Crossover

Genetic algorithm (GA) is an evolutionary optimization algorithm that mimics the processes of natural evolution in order to generate useful solution to search problems. It includes three important operations of selection, crossover, and mutation as discussed in Chapter 3.

In the course of evolution, a proportion of the population is selected to generate a new generation. Selection is the stage to choose a part of agents based on fitness value and breeds a new generation. The agents with higher fitness value called an excellent individual have higher genetic probability, and the agents with lower fitness have lower genetic probability. GA has many methods to select. Crossover is a process of exchanging subparts of two chromosomes and rearranging the order to form a new individual according to certain crossover probability. Many crossover methods exist in the GA, such as one-point crossover, two-point crossover, uniform crossover, and so on. Details have been provided in Chapter 3.

In this algorithm, the selection method is used that half of glowworms with higher fitness value is directly selected to the next generation during the course of iteration. The other half of glowworms with lower fitness value executes crossover operation to generate a new generation, and compares the fitness with the previous generation, and select half of agents with higher fitness value to the next iteration. The operation of selection and crossover can enhance the diversity of agents, avoid trapping into local optimum, and improve the convergence performance.

14.2.3.4 Hybrid Artificial Glowworm Swarm Optimization Algorithm

The detailed procedure of HGSO algorithm is as follows [4]:

Step 1: Determine the parameters of algorithm, the number of subgroups, and hierarchy, a two-layer structure model is adopted in this chapter.

Step 2: Divide the search space into j parts, partition of swarm into j subgroup, and the glowworm number of each subgroup is n/j, then initialize the position of glowworms of each subgroup.

Step 3: Search the peaks of each subgroup by GSO algorithm based on chaotic map.

Step 4: Combine the optimum of the underlying subgroup to form top glowworm swarm and find the peaks of multimodal function by the GSO algorithm. The algorithm is incorporated in the operation of selection and crossover and step size is changed.

14.2.4 Particle Glowworm Swarm Optimization

The basic GSO algorithm has shown some weaknesses in global and high-dimensional search such as slow convergence rate and low computational accuracy. The slow convergence rate is caused mainly by the individual location update model of the original GSO, which only contains local information. It is obvious that a glowworm individual changes its position only *via* local information, which slows the global convergence rate. The low computational accuracy is because of the individual updating its location in all dimensions at the same time. Sometimes, the objective function value is likely to deteriorate in a high-dimensional space if independent variables change their value in all dimensions at the same time. Inspired by the PSO, we proposed an individual particle glowworm swarm optimization (PGSO) location update according to Equations 14.11 and 14.12. According to Equation 14.11, individual location updates relate to both local and global information. In addition, the update dimensions of individual locations decrease gradually along with the increase in generations.

$$x_i(t+1) = x_i(t) + c1 * \text{rand} * [x_j(t) - x_i(t)] + c2 * \text{rand} * [x_{gb}(t) - x_i(t)] \tag{14.11}$$

where $x_i(t)$ is the location of individual i at the tth iteration, $x_j(t)$ is the location of the i's neighbor that is selected, $x_{gb}(t)$ is the location of the global optimal individual, $c1$ and $c2$ denote the acceleration factors, and rand is a random number [5].

$$m(t) = D - \text{ceil}\left(\frac{D*t}{\text{gen}}\right) + 1 \qquad (14.12)$$

where $m(t)$ is the update dimensions of location at the tth iteration, D is the problem dimension, and gen is the maximum evolution generations.

14.2.4.1 Parallel Hybrid Mutation

The detailed process of parallel hybrid mutation is given as follows [5]:

1. Solve Equation 14.13 to set the mutation capability value of each individual,

$$mc_i = 0.05 + 0.45 \frac{\left[\exp\left(5(i-1)/(n-1)\right) - 1\right]}{\exp(5) - 1} \qquad (14.13)$$

   where n denotes the population size, i denotes the individual serial number, and mc_i denotes the mutation capability value of an individual, which is represented by i.

2. Choose the mode of mutation. p_u is the mutation factor that denotes the ratio of the uniform distribution mutation; correspondingly, $1 - p_u$ is the ratio of the Gaussian distribution mutation. The Gaussian (σ) returns a random number drawn from a Gaussian distribution with a standard deviation σ. Ceil (p) generates the elements of p to the nearest integers greater than or equal to p. Here, we adopt the linear mutation factor, and the function is given as

$$p_u(t) = 1 - \frac{t}{\text{gen}} \qquad (14.14)$$

   where t denotes the current generation and gen denotes the maximum generation.

14.2.4.2 Local Searching Strategy

A modified GSO algorithm is now presented, which was inspired by the glowworms' mating behavior [5]. Enlightened by this idea, we introduced the strategy of local searching in PGSO. The strategy of local searching is somewhat similar to the above-mentioned idea but is not the same. In PGSO, we implement local search in the global optimal individual vicinity of each five generations. For each local search, we implement a search near the global optimal individual. If the objective value of a new position is better than the original position, the global optimal individual moves to the new position and the search is terminated.

Local search can be performed four times at most. Algorithm 3 is the local search algorithm.

$$x'_{gb}(t) = st(t) * \text{rand} * x_{gb}(t) \qquad (14.15)$$

where $x_{gb}(t)$ denotes the global optimal position of the tth generation, $st(t)$ denotes the step size of the tth iteration, and $x_{gb}(t)$ denotes the new position. The value of $st(t)$ is calculated *via*

$$st = st(0) * \left(\frac{1-t}{\text{gen}}\right) + 10^{-4} \qquad (14.16)$$

where $st(0)$ is the initial step size, t is the current number of iterations, gen is the maximum number of iterations, and 10^{-4} is the lower bound of the step size. The PGSO algorithm pseudocode is as follows [5]:

14.2.4.3 Particle Glowworm Swarm Optimization Algorithm

Step 1: Initialization: Set the generation $G = 1$; problem dimensions $= m$; population size $= n$; step size $= st(0)$; initial luciferin $= l_0$; initialization parameter β; γ and r_0

Step 2: Glowworms distribute: Glowworms are randomly uniformity distributed in search space. All glowworms carry an equal quantity of luciferin l_0 and own initial neighborhood domain radius r_0

Step 3: While $G <$max *generation* do

3.1: For $i = 1: n$ (all glowworms) do

Update luciferin according to Equation 14.1

Confirm the set of neighbors according to $j \in N_i(t)$, $N_i(t) = \{j:d_{ij}(t) < r_d^i(t); l_i(t) < l_j(t)\}$

Compute the probability of movement according to Equation 14.2;

Select a neighbor j using a probabilistic mechanism;

Glowworm i move toward j according to Equation 14.3;

Update neighborhood range according to Equation 14.4;

end for

3.2: If $(G\%5 == 0)$

Execution Algorithm 2 (given below)

Execution Algorithm 3 (given below)

Algorithm 2. Choose the Mutation model.

```
For i = 1: n
    If ceil(mc_i + rand − 1) == 1
    If rand <= p_u
        X_i(t) = (1 + rand) * x_i(t)
    Else
        X_i(t) = Gaussian(σ) * x_i(t)
    End if
End
```

Algorithm 3. Local Search.

```
For i =: 4
    x'_gb(t) = st(t)*rand*x_gb(t)
    If f(x'_gb(t)) < f(x_gb(t))
    x_gb(t) = x'_gb(t)
    Break
    End if
End for
```

End if

Step 4: End while

Step 5: Output and algorithm end

14.2.5 Glowworm Swarm Optimization Algorithm-Based Tribes

Glowworm swarm optimization algorithm-based tribes (TGSO) [6] is an improvement over the basic GSO by making them imitate the organizational behavior of human community specially tribes. Tribes are present in grassland communities away from the city. A tribe contains small number of people and they elect a tribal leader that shows strong ability for leadership. Next section explains how this behavior is applied to the glowworm's behavior in order to obtain better results than the conventional GSO.

14.2.5.1 Tribal Structure

Let us assume that there are in total $l \times m$ glowworms in the total swarm population P. In the first step, we divide the population P into l subpopulations. These subpopulations are denoted as tribes. The tribes show the same structure as that of the GSO model: they have m glowworms from which the t_{best} chosen have the best fitness value. The glowworms in each tribe run the GSO algorithm and chose a t_{best}, thus each tribe i ($i = 1, 2, ..., l$) has a tribal best t_{besti}. We get the global best out of the tribal best values. Figure 14.3 shows the structure of the tribes in the population. TGSO algorithm consists of two layers: the first layer is given to the tribes and the glowwormshaving the highest luciferin are given the second layer; thus, the global optima are obtained from the second layer according to GSO.

The tribes of the first layer show independent behavior in which each tribe does not do any information transfer with the other. The tribes then communicate with others using the glowworms having the highest luciferin in the second layer, in order to obtain the global optima (Figure 14.3) [6].

14.2.5.2 The Glowworm Swarm Optimization Algorithm-Based Tribes

The working of TGSO algorithm can be explained using the following steps:

Step 1: Initialize $l \times m$ glowworms, dimension m, population P, step size st, and other constant parameters.

Step 2: Divide population P into many subpopulations. Let there are l subpopulations with each having size m. Denote each subpopulation as a *tribe*.

Step 3: Initialize the glowworms in the search space randomly.

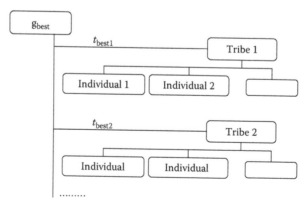

FIGURE 14.3
Layer structure of TGSO algorithm.

Step 4: Use Equation 14.1 and put the value of $J(x_i(t))$ in $l_i(t)$, where $l_i(t)$ denotes the luciferin level of glowworm i during time t and $J(x_i(t))$ denotes the cost of the objective function at glowworm i's location during time t.

Step 5: The glowworms present in the tribes selects a neighbor that has a larger luciferin value from their own in order to make up the $N_i(t)$.

Step 6: For each glowworm i in the tribe:

Choose a suitable neighbor using Equation 14.2 in the neighborhood and update its own position toward the neighbor according to Equation 14.3.

Step 7: For each glowworm i in the tribe:

Use Equation 14.4 to update the variable neighborhood range of the glowworm.

Step 8: Choose the glowworm with the highest luciferin for each tribe during time t and make the second level of the structure using them.

Step 9: The second level GSO then uses the basic GSO to choose the glowworm with the highest luciferin level.

Step 10: If number of iterations<max_iterations

        Return to step 4

    Else

        Move to step 11.

Step 11: Show the results. End.

Tribe-GSO algorithm flowchart is shown in Figure 14.4 [6].

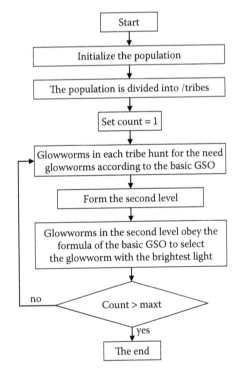

FIGURE 14.4
Flowchart for TGSO algorithm.

14.2.6 Adaptive Neighborhood Search's Discrete Glowworm Swarm Optimization Algorithm

This variant of GSO keeps a balance between the exploration and exploitation properties of any optimization algorithm. The adaptive neighborhood search's discrete glowworm swarm optimization (ADGSO) algorithm exploits the local routing solutions using local search heuristics as well as incorporates global evolutionary optimization [7]. In ADGSO, the strategy allows disregarding regions of the solution space, which do not show promising results during the algorithm execution and begins exploration of new regions. In the coming sections, a description of the concepts for the ADGSO algorithm is presented.

14.2.6.1 Adaptive Neighborhood Search's Discrete Glowworm Swarm Optimization Algorithm and Its Application to Travelling Salesman Problem

ADGSO guides the search of solutions toward more promising regions in the space of solutions. It uses a local change strategy combining different local search heuristics resulting in improved results. Generally, standard discrete optimization algorithms use the Roulette method for a discrete solution to move to new positions, but sometimes it results in retaining existing populations and encodes information quite a lot, and ultimately results in the solution falling into the local optimum or outside the scope of the search space. In order to solve these problems, ADGSO uses the mutation operation of the GA, and add four local optimization operators in the solution. These four operators are exchange mutation, inversion mutation, locus mutation, and insertion mutation. The implementation of these operators can be explained as follows [7]:

1. *Exchange mutation operator*: Two integers are initialized randomly, and exchange codes with these numbers are started. We define the mathematical operation as
   For $X = (x_1, x_2, ..., x_n)$, u and v ($1 \leq u < v \leq n$),

$$f(X, u, v) = (x_1, x_2, ..., x_{u-1}, x_v, x_u, ..., x_{v-1}, x_{v+1}, ..., x_n) \qquad (14.17)$$

   where X defines the path encoding, u and v are two random numbers ($u < v$).

   After applying exchange mutation operator, the new path encoding is denoted as $f(X, u, v)$.

   For example: Let us define two numbers randomly, $u = 2$ and $v = 5$, the old path encoding is

$$X = 3\ \underline{2}\ 1\ 9\ \underline{4}\ 8\ 7\ 6\ 5$$

   After operation, the new path encoding is

$$X = 3\ 4\ 1\ 9\ 2\ 8\ 7\ 6\ 5$$

2. *Insertion mutation operator*: Two integers are initialized randomly and then one is inserted before another. We define the mathematical operation as
   For $X = (x_1, x_2, ..., x_n)$, u and v ($1 \leq u < v \leq n$),

$$f(X, u, v) = (x_1, x_2, ..., x_{u-1}, x_v, x_u, ..., x_{v-1}, x_{v+1}, ..., x_n) \qquad (14.18)$$

where X defines the path encoding, u and v are two random numbers ($u < v$).

After applying insertion mutation operator, the new path encoding is denoted as $f(X, u, v)$.

For example: Let us define two numbers randomly, $u = 2$ and $v = 5$, the old path encoding is

$$X = 3\ 2\ 1\ 9\ 4\ 8\ 7\ 6\ 5$$

After the application, the changed path encoding is

$$X = 3\ 4\ 2\ 1\ 9\ 8\ 7\ 6\ 5$$

3. *Inversion mutation operator*: Two integers are initialized randomly and the encoding between them is reversed. The mathematical model of the operator as follows: For $X = (x_1, x_2, \ldots, x_n)$, u and v ($1 \le u < v \le n$),

$$f(X, u, v) = (x_1, x_2, \ldots, x_{u-1}, x_v, x_{v-1}, \ldots, x_{u+1}, x_v, x_{v+1}, \ldots, x_n) \qquad (14.19)$$

where X defines the path encoding, u and v are two random numbers ($u < v$).

After applying the inversion mutation operator, the new path encoding is denoted as $f(X, u, v)$.

For example: Let us define two numbers randomly, $u = 2$ and $v = 5$, the old path encoding is

$$X = 3\ 2\ 1\ 9\ 4\ 8\ 7\ 6\ 5$$

After the application, the changed path encoding is

$$X = 3\ 7\ 8\ 4\ 9\ 1\ 2\ 6\ 5$$

4. *Locus mutation operator*: Two integers are initialized randomly and the encoding of a number is exchanged with the one present in front of it. The mathematical model of the operator is as follows: For $X = (x_1, x_2, \cdots, x_n)$, u and v ($1 \le u < v \le n$),

$$f(X, u, v) = (x_1, x_2, \ldots, x_{u-1}, x_v, x_{v-1}, \ldots, x_{u+1}, x_v, x_{v+1}, \ldots, x_n) \qquad (14.20)$$

where X defines the path encoding, u and v are two random numbers ($u < v$).

After applying locus mutation operator, the new path encoding is denoted as $f(X, u, v)$.

For example: Let us define two numbers randomly $u = 2$ and $v = 5$, the old path encoding is

$$X = 3\ 2\ 1\ 9\ 4\ 8\ 7\ 6\ 5$$

After the application, the changed path encoding is

$$X = 3\ 1\ 2\ 9\ 4\ 8\ 7\ 6\ 5$$

14.2.6.2 Some Other Features of the Algorithm

1. *The import of the bulletin board*: In discrete solution algorithms the best solutions, which are generated in the process of iteration might deteriorate in the later iterations, which causes the algorithm to give bad results [7]. Thus, ADGSO algorithm overcomes these limitations by using a bulletin board to record the best state. It is implemented as a function, which compares the quality of the glowworms from which the optimum one is registered in the bulletin board. If the new solution is better, it is replaced with the one in the bulletin board; otherwise, it is left in order to guarantee that the best possible solution remains in the bulletin board.

2. *Confirm the departure city*: We might encounter a problem of similar paths in the encoding. For example: the paths 1-2-4-3-5-8-7-6-9 and 7-6-9-7-6-9-3-5-8 look different in formation; however, they represent the same formation of intermediate nodes as a loop that starts from the node 1, and then proceeds to the node 2, 4, 3, 5, 8, 7, 6, 9 one time each and then returns to node 1. Thus to avoid this complication, we set the node 1 as the starting node, that is, fix the location of the starting node. As the initial node is fixed, it makes this coding different, that is, the path 1-2-4-3-5-8-7-6-9 and 3-5-8-7-6-9-1-2-4 will be considered as same.

3. *The establishment of the initial population*: For initializing the population instead, choosing a sequence we choose a random arrangement from numbers 1 to n, and then repeat the step M times where M being the size of the population. Thus, the solution obtained through this procedure is all feasible and the diversity of the population is ensured.

4. *Repair unfeasible code*: The unfeasible code should be repaired as there might be duplication of codes after updating it. Here, if we find the length of code less than 1, we set it to 1; otherwise, set it to n and then find the duplicate code. If any duplicate code is found, replace it with a coding that has not appeared before.

14.2.6.3 Adaptive Neighborhood Search's Discrete Glowworm Swarm Optimization Algorithm Steps

The steps for implementing the ADGSO algorithm can be explained as follows [7]:

Step 1: Initialize the glowworm swarm population P and build the population on the search space. Initialize the parameters like population size as gsonum, maximum iterations as iter_max, initial fluorescein value as $l_i(0)$, initial variable neighborhood range as $r_i^d(0)$, maximum radius as r_s, fluorescein update value as γ, fluorescein volatile coefficient as ρ, neighborhood threshold as n_t, radius update rate as β, and parameters for update formula as $p1$ and $p2$.

Step 2: For an individual glowworm i, calculate the value of the objective function, then calculate the fluorescein value using the objective function.

Step 3: Determine the distance between the glowworm i and glowworm j, and then considering its neighborhood radius $(r_i^d(t))$, pick glowworm whose fluorescein value is higher than itself from its neighbor set $(Ni(t))$.

Step 4: Assign the probability of glowworm i to move to the glowworm $j(j \in Ni(t))$, and then using the Roulette wheel select the object in motion; In case neighbor set $(N_i(t))$ is empty, it clearly shows that glowworm i is in a local optimum, go to Step 6.

Step 5: Form a sequence r randomly such that, $r = (r_1, ..., r_k, ..., r_n)$, where $r_i \in [0,1]$. Modify the glowworm's code according to the dimension specified.

Step 6: Choose a random equation from Equations 14.17 through 14.20 to perform neighborhood search.

Step 7: Recalculate the glowworm's neighborhood radius.

Step 8: If the number of iterations of the algorithm have reached the max number of iteration or if the termination condition is reached then end; otherwise, move to Step 2, and set $i = i + 1$, continue to execute the algorithm. If the end is reached, output the global optimal solution and optimal path.

14.3 Convergence Analysis of Glowworm Swarm Optimization Algorithm

GSO algorithm is especially useful in finding multimodal solutions in a search space. It performs considerably better than most of the traditional algorithms in finding multiple peaks. GSO has the capability to find large number of peaks as well as it can also tackle discontinuities in the objective function. As the dimension of the problem increases, more glowworms are required for finding optimal solutions, but the number of glowworms that we require for some dimensionality varies according to factors such as nature of the peaks, sensor range, and so on. We see that the peaks, which are on the interior of the function, are easier to locate than on the corners. As the dimensionality of the problem increases, these effects get less observable. The experimental parameters also affect the performance of the problem. The quantities l_0, n_t, s, β, ρ, and γ are algorithm parameters for which extensive experimentation is done in order to derive suitable values for them. GSO algorithm shares some similarities with the PSO algorithm like in both algorithms a group of agents are first deployed on the solution space according to the objective function. Then each agent changes its position according to a position update rule. In PSO, the position of the particle is affected according to its own global and local best value in the solution space, whereas GSO movement is according to the neighbors in a predefined range of any particle.

With experiments, it is observed that the GSO algorithms perform better than niche PSO: a variant of PSO for finding multimodal solutions. The popularity of PSO algorithm and the superiority of GSO over PSO create a strong stance in favor of GSO algorithm's performance in complex optimization problems.

14.4 Applications of Glowworm Swarm Optimization Algorithms

The GSO algorithm has been used in various applications of the real world especially the ones requiring the multipeak solutions. A few of such applications will be discussed in the sections below. The algorithms applied to the problem are GSO or its variant according to the problem's area of application.

14.4.1 Hybrid Artificial Glowworm Swarm Optimization for Solving Multidimensional 0/1 Knapsack Problem

The most important step for solving the 0/1 knapsack problem using GSO was to properly implement the encoding of the parameters. We define the encoding process as follows [8]:

Let x_i^t be the position of glowworm at time t and $x_i^t = [x_{i1}^t, x_{i2}^t, x_{i3}^t, \ldots x_{id}^t]$ (let the number of dimensions and quantity of loading knapsack be d). Each glowworm corresponds to a solution in the search space. If it is found that $x_{ij}^t = 1$, ith glowworm puts the jth object into the knapsack.

If it is not the case, then do not put the object. The algorithm will be defined as follows, where various parameters during initialization are: let n be the glowworms count, l_0 be the starting luciferin value, r_0 be the variable neighborhood range, b be the update rate of the variable neighborhood range, n_t be the neighborhood threshold, ρ be the dissipation rate of luciferin, iter_max be the maximum number of iteration, initial iteration $t = 1$, and s be the step size.

Step 1: Initialize all the parameters and the glowworm population P in the search space. Glowworm initial position can be indicated through a random initial matrix of 0/1 defined as

$$X_{ij} = \begin{cases} 1 & \text{rand} > 0.5 \\ 0 & \text{rand} < 0.5 \end{cases} \tag{14.21}$$

Here 0.5 is set as the boundary of the discrete function, as it being middle of the two possible states (0,1).

Step 2: Use the known greedy strategy to solve the knapsack problem.

Step 3: Find the fitness value of the following solution.

$$f(x_1, x_2, \ldots x_m) = \sum_{j-1}^{n} x_j c_j, \ s.t \sum_{j-1}^{n} x_j a_{ij} \leq b_i, \ i = 1, 2, \ldots g \tag{14.22}$$

Step 4: Use Equation 14.1–14.4 to update the luciferin parameter, variable neighborhood range, neighborhood threshold, and the position of glowworms.

Step 5: If iteration_number < iter_max, return to Step 4; otherwise, terminate and show the output.

14.4.2 Glowworm Swarm Optimization Algorithm for K-Means Clustering

A glowworm swarm variant was used for image classification known as Image classification GSO (ICGSO) and performed successful K-means clustering on the image [9].

The procedure for the application of ICGSO algorithm is as follows:

Step 1: Initialize the parameters of the algorithm, which are required in computations.

Step 2: Glowworms are initialized, so as to have exactly N_c cluster centers selected randomly.

Step 3: For glowworm x_i

3.1: For pixel m_p

Computed (m_p, z_{ij}) for all C_{ij}

Set m_p to C_{ij}

$$d(m_p, z_{ij}) = \min_{\forall c=1,\dots,N_c}\{d(m_p, z_{ic})\} \tag{14.23}$$

Compute the fitness $f(x_i, M)$

Step 4: For iter $= 1$ to iter$_{max}$,

For each glowworm x_i

Updating luciferin value;

Selects conforms to the condition glowworm according to equation;

The K-means algorithm is applied to GSO of each iteration.

4.1: For each pixel m_p

Calculate for all C_{ij} using Equations (7); Assign m_p to C_{ij} where

4.2: Assign new cluster means;

Compute the fitness of the cluster

Recalculate the search radius

Step 5: Perform segmentation of the image using best cluster centers obtained using the GSO algorithm.

14.4.3 Discrete Glowworm Swarm Optimization Algorithm for Finding Shortest Paths Using Dijkstra Algorithm and Genetic Operators

This problem of shortest path is solved using the DGSO algorithm [10]. Here, the GSO algorithm is used along with genetic operators and the Dijkstra algorithm in order to find a shortest path from the source to destination in a graph. Several techniques, which are used in the basic GSO algorithm in order to solve the shortest path problem effectively, are explained as follows:

14.4.3.1 Labeling Method

A binary format of coding is used in the algorithm. For increasing the performance of the algorithm, an optimized initial solution is chosen and some constraints are added. In order to prevent a loop in the initial population, we chose the next point after the starting point randomly from the point attached from the starting point again and again, till the start point is reached again. For preventing further middle nodes to have repetitive values, a labeling method is used to label the nodes in the Dijkstra algorithm, so only nodes that are not labeled initially can be used for selecting the new path node. If node breaks in the middle, program get restarted.

14.4.3.2 Roulette Selection Strategy

GA uses this strategy as the selection method in their reproduction stage as discussed in Chapter 3. In this strategy, an individual being selected for an operation is directly proportional to the ratio of the individual fitness value to the fitness of the total population. So technically, higher fitness of an individual results in more chance for an individual to get selected.

The algorithm uses this strategy in order to select glowworms that are able to diversify other glowworms' positions and prevent them from converging suboptimally.

14.4.3.3 Single-Point Crossover Strategy

It refers to a random location selection from the individual coding string, which results in the transfer of genes after that point among the two individuals. The steps for single point crossover can be given as follows [10]:

- Glowworm j is determined using the Roulette strategy.
- Nonzero genes are to be found between glowworm i and j, that is, the same nodes presents in the two paths, and a random gene is chosen as the point of crossover to ensure feasible solutions after there is exchange among them.
- The gene after the point of crossover for glowworm j are to be given to glowworm i, so as to make a new glowworm while keeping glowworm j unchanged.
- Give the gene after the crossover point of j to the corresponding position of i to produce a new glowworm, and keep the genes of j unchanged.

14.4.3.4 Mutation Strategy

Mutation is done to change the genes of population randomly in order to explore new solutions and to have diversity in solution space. It makes the particles move out of local optima, in order to find the global optima. The paths which do not have high fitness are mutated to form different solutions, which might show better behavior. The process of mutation can be explained as follows [10]:

- All the glowworms are to be sorted according to the fitness value.
- Find the solution having the worst fitness value, and then select a random position in that solution for mutation.
- After the path is mutated, all the genes are regenerated according to the labeling method to obtain feasible solutions.

14.4.3.5 Procedure of Glowworm Swarm Optimization Algorithm for Finding Shortest Paths

GSO algorithm for shortest path finding can be summarized as follows [10]:

Step 1: Initialize population P. Use labeling procedure as given in Section 14.4.3.1, then initialize the parameters of glowworms used for decision-making.

Step 2: Identify the solutions with the worst fitness value and do mutation of the solutions.

Step 3: Use Equation 14.1 to calculate the fluorescein value of the glowworms.

Step 4: Identify the movement of the glowworms using the Roulette strategy and Equation 14.2.

Step 5: Make changes in the location using single point crossovers and then update the radius of decision domain using Equation 14.3.

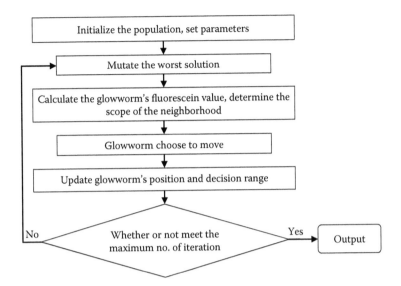

FIGURE 14.5
Flowchart of finding shortest paths using DGSO.

Step 6: If the maximum iterations are completed, terminate the algorithm and output the results; otherwise, move to Step 2. End.

The detailed flowchart of the procedure can be given as shown in Figure 14.5 [10].

14.5 Search Capability of Glowworm Swarm Optimization Algorithm

The algorithm is applicable to a variety of problems; however, its major speciality is for finding multimodal solutions. The algorithm is presented for maximization problems in the original literature. However, with modifications it is shown to give good results for finding multiple minima for multimodal functions. The algorithm shows better results to find multimodal solutions than other traditional algorithms. When defining GSO at the individual level: two group level phases are observed. The two group level implementations are formations of dynamic networks, which give rise to splitting of the swarm into subswarms and in each subswarms obtained the glowworms tend to get into local optima.

14.6 Summary

In this chapter, we discussed GSO and some of its variants and how they can be applied to various problems of real life. The algorithm, however, being similar to PSO in basic implementation, shows several notable differences, which make it totally different from the former. The most important areas of application are those which require multimodal solutions for their problem. GSO algorithm results in computation of multiple optima in parallel within a single program run, and this provides alternative solutions for the user to consider. The result of multiple solutions is sometimes beneficial in cases where some solutions are more feasible to implement in real life than the other. The performance of the algorithm also depends on several factors such as the dimensions, initial parameters, and so on. The parameter initialization for the search space and so on are beyond discussion and require some monitoring and experimentation and vary highly from case-to-case depending on the situation and data bound of the considered problem, whereas the higher dimensionality requires more number of solution particles in order to efficiently detect the solutions. The major variants of the GSO discussed in this chapter are as follows:

- The traditional GSO algorithm.
- The variants of the GSO algorithm showing better exploration and exploitation of solution particles according to the problem statement or in general. Some specific variants discussed in this chapter are as follows:
  - Hybrid coevolutionary GSO.
  - GSO with random disturbance factor (R-GSO).
  - GSO algorithm based on hierarchical multisubgroup.
  - PGSO.
  - TGSO.
  - ADGSO algorithm.
- The discussion related to the convergence and dynamics of the GSO Algorithm and how the different variants are proven to be better.
- A few examples to show how the GSO algorithm is working both with respect to continuous and discrete problems.

Solved Examples

1. What makes GSO useful for finding simultaneous solutions of multimodal functions?

   Ans: GSO algorithm uses the concept of variable neighborhood range for determining the movement of glowworm agents in the next iterations. This variable neighborhood range makes the particles to form disjoint subgroups in which individual particles are not affected by other particles, which are outside their neighborhood range. This makes the algorithm's solution space to form various optimal clusters based on the strongest members in a specific range of solutions, and thus multiple solutions can be effectively obtained using the GSO algorithm.

2. Explain the GSO algorithm working on an optimization of a multimodal function?

   Ans: The GSO algorithms are a nature-inspired meta-heuristic algorithm working on the natural behavior of glowworms. This algorithm is found successfully to be able to find solutions for multipeak problems.

In order to understand the GSO algorithm, refer Sections 14.1.1 and 14.1.2.

We try to prove that the algorithm works for solving the multimodal solutions by trying to implement it on the *Rastrigin's function*. Rastrigin function is a standard benchmark function used for testing optimization algorithms. It is a multimodal function with multiple simultaneous solutions to be obtained in the function. The standard equation of this function is presented as follows (Figure 14.6):

$$f(x) = 10N + \sum_{i=1}^{N} x_i^2 - 10\cos(2\pi x_i) \tag{14.26}$$

The simulations of the experiment were coded using MATLAB® 7.0 and implemented using machine with 2 GB RAM, 3 GHz Intel P4 Machine, and a Windows XP Operating System [2]. The algorithm is applied on the function optimization problem in order to obtain the solutions and it was found that the algorithm worked successfully in finding the multiple peaks of the function. A diagram for the capture of the Rastrigin function solutions is presented as in Figure 14.7 [2].

The diagram is shown in four phases. The phase (a) defines the initial position of the glowworm swarm. In phase (b) the glowworms are seen to be moving toward the best particle in their neighborhood range. In phase (c) the converged state is shown when the neighborhood is kept constant and phase (d) shows the converged state when the neighborhood is kept variable.

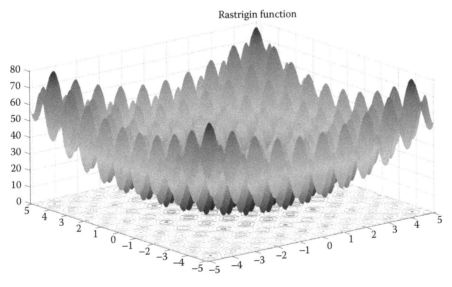

FIGURE 14.6
Rastrigin Function.

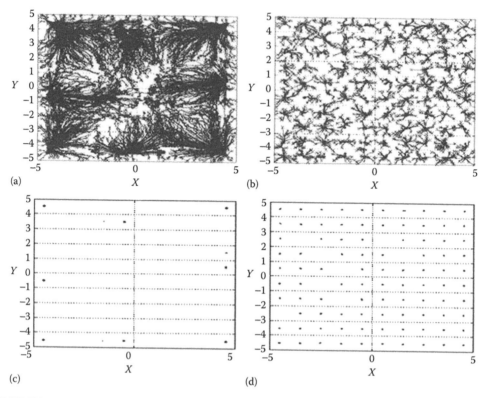

FIGURE 14.7
The convergence of glowworms for Rastrigin function. (a) Initial position glowworm swarm; (b) Movement of glowworm swarm; (c) Converge state when neighborhood is constant; (d) Converge state when neighborhood is variable. (From Krishnanand, K. N. and Ghose, D., *Swarm Intell.*, 3, 2, 87–124, 2009.)

Exercises

1. Explain the three phases of implementation in the GSO algorithm.
2. Implement GSO algorithm in a programming language of your choice.
3. Comment on the usability of GSO algorithms. What are the set of problems over which glowworm swarm is especially advantageous?
4. Explain some similarities and differences between PSO and GSO. What is the variant of PSO for finding multimodal solutions?
5. Discuss various parameters used in GSO algorithm and their importance.
6. Discuss any one variant of GSO algorithm.
7. How is Roulette selection strategy employed for DGSO?
8. Describe other nature-inspired algorithms or its variants, which are used to find multimodal solutions?
9. Use DGSO algorithm for solving Traveling salesman problem: 10 Cities, that is, A – J alphabetically, are located at the Cartesian locations (10,0), (20,5), (35,10), (40,15), (25,20), (25,30), (20,30), (10,20), (5,5) and (25,10), respectively.

a. Solve the above TSP for Particle Count = 8 if maximum velocity changes are bound to 5. You may choose a convenient upper bound for the maximum number of iterations for the algorithm.

b. Solve the above-mentioned TSP for Particle Count = 4, if the maximum velocity changes are bound to 10. You may choose a convenient upper bound for the maximum number of iterations for the algorithm.

c. Comment on the performance of the algorithm in the above-mentioned two cases, that is, 5a and 5b.

10. Compare the performance of the results obtained in the above-mentioned problem 9 to that of the results obtained after the application of PSO algorithm.

References

1. K. N. Krishnanand and D. Ghose, Glowworm swarm optimisation: A new method for optimising multi-modal functions, *International Journal of Computational Intelligence Studies*, 2009, 1, 1, 93–119.
2. K. N. Krishnanand and D. Ghose, Glowworm swarm optimization for simultaneous capture of multiple local optima of multimodal functions, *Swarm Intelligence*, 2009, 3, 2, 87–124.
3. Q. Liangdong, D. He and J. Wu, Hybrid coevolutionary glowworm swarm optimization algorithm with simplex search method for system of nonlinear equations, *Journal of Information & Computational Science*, 2011, 8, 13, 2693–2701.
4. L. He, X. Tong and S. Huang, Glowworm swarm optimization algorithm based on hierarchical multi-subgroups, *Journal of Information and Computational Science*, 2013, 10, 1245–1251.
5. T. Zhonghua and Y. Zhou, A glowworm swarm optimization algorithm for uninhabited combat air vehicle path planning, *Journal of Intelligent Systems*, 2015, 24, 1, 69–83.
6. Z. Yongquan et al., A glowworm swarm optimization algorithm based tribes, *Applied Mathematics & Information Sciences*, 2013, 7, 2L, 537–541.
7. D. Wenbo et al., *Bio-Inspired Computing-Theories and Applications*, Springer, Berlin, Germany, 2015.
8. G. Qiaoqiao, Y. Zhou and Q. Luo, Hybrid artificial glowworm swarm optimization algorithm for solving multi-dimensional knapsack problem, *Procedia Engineering*, 2011, 15, 2880–2884.
9. Z. Yongquan et al., A novel K-means image clustering algorithm based on glowworm swarm optimization, *Przegląd Elektrotechniczny*, 2012, 88, 8, 276–270.
10. Z. Xin, S. Chen and Y. Sheng, Research on the problem of the shortest path based on the glowworm swarm optimization algorithm, *15th COTA International Conference of Transportation Professionals*, Beijing, China, July 24–27, 2015.

15

Bacteria Foraging Optimization Algorithm

15.1 Introduction

These days, the term efficiency is fundamentally present in each specialist's vocabulary. For example, execution and cost cannot be disregarded in an aggressive society such as our own. Minimizing cost as well as performance maximization can be considered as an improvement issue so that, to enhance is to locate the best answer for a specific assigned issue. Each method has an arrangement of issues for which it is more specific. This relies on the progression of issue qualities, uncommonly the capacity depicting it, and is not effectively reachable. Consequently, a great general comprehension of the issue and also of the optimization technique is required.

Techniques that are deterministic depend on the functions, their derivatives, or the approximations of it. The objective then is to discover where the slope of the function is null using the direction of the slope to where it is headed. Nondeterministic techniques scan for the ideal solution of a specific issue. In a huge number of issues, there is a need to discover ideal arrangements because of more than one or many trademarks. For this situation, a multitarget approach must be utilized. Engineering issues requesting high-performance solutions and low losses are cases of use in which this kind of approach is required. In this chapter, we introduce another nature-inspired optimization technique i.e., BFOA.

15.2 Biological Inspiration

The principal motivation behind this bacterial foraging optimization algorithm (BFOA) is that its calculation is not to a great extent influenced by the size and the problem nonlinearity.

BFO algorithm has points of interest, like less computational time, less computational weight, and global convergence, and can deal with many number of given objective functions as and when they are compared with the other algorithms, which are evolutionary in nature.

For multioptimal function optimization, use of strategy which is group foraging of a swarm of *Escherichia coli* bacteria is the key thought of the new proposed algorithm. The search task for the nutrients by the bacteria is done in a way which amplifies the energy received per unit time. Singular bacterium likewise speaks with other bacteria by sending various signals. A bacterium will take foraging choices by taking into account the last two factors. The procedure, in which a bacterium while hunting down and taking nutrients moves by making little steps and strides, is called by the name chemotaxis and the main thought of BFOA is impersonating chemotactic development of some given virtual bacteria in the problem space.

Since its initiation, BFOA has caught the consideration of specialists from different areas of learning, particularly because of its natural inspiration and structure. Specialists are attempting to combine BFOA with various different algorithms, so as to investigate its nearby and global pursuit properties independently. It has as of now been connected to numerous true issues and demonstrated its viability over numerous variations of GA and PSO.

15.3 Bacterial Foraging Optimization Algorithm

Amid the foraging process of the genuine bacteria, locomotion of the bacteria is accomplished by an arrangement of elastic flagella. The tumbling or swimming of *E. coli* bacteria is caused by the flagella which, at the time of foraging [1,2], are two essential locomotion operations performed by a given bacterium. All the flagella pull on the cell, when turned in the clockwise direction. That outcome in the smoother moving of flagella autonomously and finally, the bacterium tumbles with lesser tumbling, whereas in a hurtful place it will tumble every now and again to locate a nutrient slope. Movement of the given flagella in the opposite direction helps the *E. coli* bacterium to make swimming at a quick rate. In mentioned earlier, the *E. coli* bacteria experience a process called chemotaxis, in which they get a kick out of the chance of moving toward a nutrient slope and stay away from poisonous environment. For the most part, the bacteria move for a more drawn-out separation in a well-disposed environment. Virtual bacterium movement on the functional surface to locate the global optimum is shown in Figure 15.1 [3].

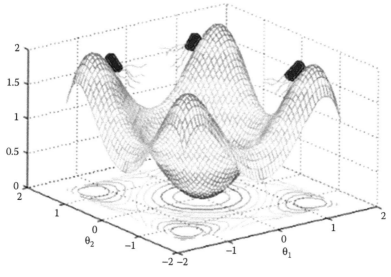

FIGURE 15.1
An objective function which has multimodal surface along with a bacterial swarm.

When they will get nourishment in adequate, they are expanded long and in nearness of appropriate temperature they get soften up in the center to form a correct copy of themselves. On account of the event of abrupt ecological change and assault, the chemotactic advance might be demolished and the gathering might move to other different spots or others might be presented in the swarm or group of interest. This entire process makes the occasion of end dispersal into the real and genuine bacterial populace, where every one of the microorganisms inside a locale is slaughtered or a gathering is scattered into another part of the bacterium's environment.

Now, assume that we need to locate the minima of (J,θ) where p $\theta \in \mathfrak{R}$ (i.e., θ is created as a vector of real numbers of containing dimensions of number p), and also we do not have estimations of the gradient vector $\nabla (J,\theta)$. BFOA algorithm impersonates the four primary instruments as seen in a genuine and real bacterial framework: chemotaxis, reproduction, swarming, and elimination dispersal to take care of this optimization issue of nongradient cases. A bacterium, which is virtual is the one with really one trial arrangement (might be known as a kind of search agent), which proceeds onward the functional surface to find the given global optima.

Using the foraging technique of a group of an *E. coli* swarm microscopic type of bacteria in a multioptimal setting with optimization of function is the key thought of this newly proposed algorithm. Microscopic organisms searching for supplements are a way to increase energy acquired per given unit time. Singular bacterium likewise speaks with other bacterium by sending some kind of signal. A single bacterium will take foraging decisions using past two events or factors.

A step that is chemotactic is defined by a tumble, which is followed by a run or a tumble, which is followed by another tumble. Let the chemotactic step be defined by an index number j, k for the reproduction step and let us say l be the index for elimination–dispersal step.

BFOA copies the chemotactic movement of bacteria search space of the problem.

Let,

p be the number of dimensions of a given search space

S be the number of total bacteria in the given population

N_{ed} be the elimination dispersion type of event count

N_c be the chemotactic steps count

N_s be the swimming length

P_{ed} be the probability of elimination–dispersal

$C(i)$ be the size of steps taken in any direction, which is random specified by the tumble

Let $P(j, k, l) = \{I = 1,2,..., S \mid \theta^i (j, k, l)\}$, at the jth chemotactic step that represents the position coordinates of each member of bacteria in the given population of the S bacteria, kth reproduction step with lth elimination–dispersal event. The cost at the location of the ith bacterium be $J(i, j, k, l)$, let denote $\theta^i (j, k, l) \in \mathfrak{R}$ (sometimes ith bacterium position is referred as θ^i). S can be very large for actual bacterial populations. Using simulations, we will be using much smaller population sizes and the population size will be kept fixed. BFOA, however, can be applied to higher dimensional optimization problems.

Four prime steps are used to explain bacterial foraging optimization theory:

- Chemotaxis
- Elimination–dispersal
- Swarming
- Reproduction

15.3.1 Chemotaxis

This procedure mimics the development of an *E. coli* cell by means of flagella using swimming and tumbling. Organically in two distinctive ways, an *E. coli* bacterium is able to move. It is able to swim for a time frame in a similar course or it might tumble and exchange between these two given methods of process operation for the whole lifetime. Assume θ^i (*j*, *k*, *l*) speaks to bacterium number *i* at chemotactic step *j* with *k*th regenerative and *l*th elimination dispersal step. $C(i)$ is the extent of the progression taken in a way which is a random course indicated by the tumble. At that point in computational end of chemotaxis, the development of the bacterium might be given by

$$\theta^i(j+1,k,l) = \theta^i(j,k,l) + c(i)\frac{\Delta(i)}{\sqrt{\Delta^{\mathrm{T}}(i)\Delta(i)}} \tag{15.1}$$

where Δ is an initial vector in a direction, which is random whose elements are in $[-1,1]$.

15.3.2 Swarming

A fascinating gathering conduct has been watched for several mobile types of bacterium such as *E. coli*, where complicated swarms are shaped in a kind of nutrient medium. A gathering of *E. coli* bacteria cells orchestrate themselves in form of a voyaging ring by climbing up the nutrient slope when set in the mid of a partially solid matrix with a solitary nutrient known as chemo-effecter. Stimulation of cells with an abnormal state of succinate, discharge an attractant medium, which helps these cells to faster aggregate into gatherings and in this way move as a concentric pattern of groups or swarms with a very high density of bacteria.

The signaling pattern in the swarm of *E. coli* might be given by

$$J_{cc}[\theta, P(i,j,l)] = \sum_{i=1}^{s} J_{cc}[\theta, \theta^i(i,j,l)]$$

$$= \sum_{i=1}^{s}\left[-d_{\text{attractant}}\exp\left(-w_{\text{attractant}}\sum_{m=1}^{p}(\theta_m - \theta_m^i)^2\right)\right] \tag{15.2}$$

$$+ \sum_{i=1}^{s}\left[h_{\text{repellant}}\exp\left(-w_{\text{repellant}}\sum_{m=1}^{p}(\theta_m - \theta_m^i)^2\right)\right]$$

where J_{cc} [θ, $P(j, k, l)$] is the original function value, which is to be added to the given actual objective function that has to be minimized, S is the total bacteria count, and p is the count of variables to be optimized.

15.3.3 Reproduction

The bacterium whose health is at the minimum dies and the healthy bacterium split into two bacteria and keeps the size of the swarm steady.

15.3.4 Elimination and Dispersal

On account of different reasons, sudden or continuous changes happening in the neighboring environment where a bacteria population is living may happen. The end goal of the events could be each microorganism inside a locale is executed or sometimes a gathering is scattered into another part of the environment. For instance, a critical local rise in temperature may execute a gathering of bacteria that are at present in a locale with a high grouping of nutrient gradient. Over drawn out stretches of time, such type of events had spread different sorts of bacteria into all aspects of our surroundings from our digestion systems to hot water springs and in underground areas. To recreate this wonder in BFOA, a few microbes are exchanged indiscriminately with little likelihood considering that the new substitutions are arbitrarily introduced over the pursuit space.

Dispersal occasions have the impact of potentially wrecking chemotactic advance, yet they likewise have the impact of helping with chemotaxis and as their dispersal may put the microbes close great sustenance sources. From a wide point of view, disposal and dispersal are parts of the populace level long-separate motile conduct.

Bacterial Foraging Optimization Algorithm

The BFOA algorithm is as follows:

1. Initialize S, N_c, Ns, N_{re}, P_{ed}, N_{ed}, and the $C(i)$ where $(i = 1,2,....,S)$ parameters. Initialize θ^i, $i = 1....S$.
2. Loop for elimination–dispersal step: $l = l + 1$.
3. Loop for reproduction: $k = k + 1$.
4. Loop for chemo taxis: $j = j + 1$.
   a. Taking a chemotactic step for bacteria "i" while $i = 1, 2,....,S$:
   b. Calculate $J(i, j, k, l)$.
   c. $J(i, j, k, l) = J(i, j, k, l) + J_{cc}(\theta i \, (j, k, l), P(j, k, l))$.
   d. $J_{end} = J\,(i, j, k, l)$.
   e. Tumble: Generate a random vector $\Delta(i)$ with each element on $[-1,1]$.

$$\theta^i(j+1,k,l) = \theta^i(j,k,l) + C(i)\frac{\Delta(i)}{\sqrt{\Delta^T(i)\Delta(i)}}$$

   f. Calculate $J(i, j + 1, k, l)$.
   g. Swimming step.
   h. Let $m = 0$ (swim length counter)

   i. While $m < N_s$:

      $m = m + 1$

      If $J(i, j + 1, k, l) < J_{last}$: (improvement)

         $J_{end} = J(i, j + 1, k, l)$ and

$$\text{let } \theta^i(j+1,k,l) = \theta^i(j,k,l) + C(i)\frac{\Delta(i)}{\sqrt{\Delta^T(i)\Delta(i)}}$$

      and use this $\theta^i(j + 1, k, l)$ to compute the new $J(i, j + 1, k, l)$.

   Else:

      $m = N_s.$

   j. Go to next bacterium $(i + 1)$ if $i \neq S$.

5. If $j < N_c$ move to Step 3.

6. Step for reproduction

      $m = N_s$

   a. Given k and l let $J_{health}{}^i = \sum_{j=1}^{Nc+1} J(i,j,k,l)$ be the heath of i bacteria. In order of ascending cost J_{health}, sort bacteria and chemotactic parameter $C(i)$.

   b. Highest J_{health} valued bacteria dies and the other bacteria with the best values split.

7. If $k < N_{re}$, go to Step 2. Here the number of desired reproduction steps are not reached.

8. Elimination–dispersal

   For $i = 1,2,.....,S$ eliminate and disperse each bacterium with probability P_{ed}.

   The FVSI, L-index, power loss, total cost, and bus voltage are obtained separately.

15.4 Variants of Bacterial Foraging Optimization Algorithm with Applications

There are many variants of the BFOA that has been proposed with applications in various domains; some of them are discussed as follows:

In [4] BFO was applied to radio frequency identification (RFID) for network scheduling. A BFO variant is developed to balance the exploration/exploitation trade-off, which is called self-adaptive bacterial foraging optimization (SABFO) in which the swimming length of a bacteria changes during the time of search. Amid pursuit if any bacterium finds a promising space (improved fitness value), swimming length of microscopic organisms is adjusted to littler one (misuse state). On the off chance that a bacterium's wellness is unaltered for a user's characterized steps when food gets depleted, the swimming length acclimates to bigger one and this kind of bacteria enters in a state of exploration. When contrasted with genetic algorithms (GAs), particle swarm optimization (PSO) and bacterial foraging optimization (BFO) demonstrate that SABFO gets predominant arrangements than alternate strategies.

For short-term electric load forecast, application of bacteria foraging optimized algorithm using a neural network (BFO NN) was presented by Zhang et al. [5]. Global inquiry highlight of BFO prompts to a quicker merging of neural net. For preparing an artificial neural network (ANN), BFO has been used in some kind of feedback way. Cost function to BFO is taken as the mean square error (MSE) of the neural net. BFO is utilized to discover streamlined weights of the artificial neural system while minimizing the value of the MSE. This model was connected for load prediction of New York City and researchers got extremely precise outcomes. Recreation comes about likewise demonstrated BFO NN focalizes more rapidly than genetic algorithm upgraded neural network (GANN).

In [6] portrays the use of BFO in solving the antenna problem. The antenna problem presents utilization of BFO algorithm for streamlining of included point of V-dipole receiving wire for obtaining higher antenna directivity. The method of moments (MoM) coding for that has been combined with BFO to get the best possible angle. Directivity of the given V-dipole, which is also the capacity of included point, has to be taken as the cost capacity to be maximized. As BFO is typically used for cost minimization, a fitness value corresponding to directivity is considered as the target or as the cost function, which has to be minimized. Relative outcomes between V-dipole and straight line dipole with upgraded included point utilizing BFO algorithm have been given.

T. Datta et al. [7], exhibit an enhanced versatile approach including adaptive BFO (ABFO) to enhance both the phase and the amplitude of weights of a straight cluster of given antennas for greatest factor at any sought course and null in a particular angle. To make the swimming step size of BFO algorithm versatile, guideline of adaptive delta modulation has been utilized. The technique is connected to six components of straight cluster. It is found that ABFO is fit for combining designs with various nulls at any wanted bearings and that too with better convergence.

In [8] BFO is utilized to ascertain some kind of resonant frequency and also the local feed point of a small-scale strip reception apparatus containing substrates, which are thick (having h/λ greater than 0.0815). Changes appearing in wide widths of patch, for given estimations of a substrate length, thickness, width, and encourage the purpose of the patch, are figured utilizing some of the standard expressions and are then haphazardly set in inquiry space. At that point, improvement is always done to get the test estimation of the resonant frequency of the patch. The optimized width of the patch is then used to figure resonant frequency of the patch. To acquire bolster point area utilizing BFO, condition input impedance fix radio wire has been taken as the cost capacity of streamlining.

In [9] the authors present a crossover approach, including PSO algorithm and BFO for improvement of multimodular and very high-dimensional capacities. The calculation joins PSO-based change administrator with some kind of bacterial chemotaxis keeping in mind the end goal to make prudent utilization of investigation and misuse capacities of the given search space and to stay away from false and untimely convergence. The calculation is tried on five standard capacities Griewank, Ackley, Rastrigin, Rosenbrock, and Shekel's Foxholes furthermore on spread range radar polyphase code design. It has been found that the general execution of the hybrid method is superior to standalone BFO and in any event tantamount to PSO and its variations.

In [10] PSO is used by Korani called BFPSO, for tuning and setting of PID-type controller k_i, k_p, and k_d. Routine BFO relies on irregular inquiry bearings, which may prompt to defer in achieving global arrangement, whereas PSO is inclined to be caught in the nearby minima. With a specific end goal to show signs of improvement advancement, the new calculation joins points of interest of both the calculations that is, PSO's capacity to trade social data and BFO's capacity in finding new arrangements by disposal and dispersal. Simulations showed that overshooting cases of a PID controller are diminished extensively with speedier convergence and in this manner the cross-breed calculation beats routine PSO and BFO.

In [11] gives a mixed hybrid kind of approach, including the BFO and PSO. The creators used property of speedier convergence of PSO algorithm for finding relating places of microorganisms in the given predefined area, that is, seek space. By doing as such they made the inquiry space limited and accordingly diminishing the required computational period. The new change in full frequencies of various patches is ascertained first and set arbitrarily in the pursuit space of PSO and after that looking begins utilizing the PSO to closest test values. At that point postulation qualities are set in an inquiry search space of BFO. The root mean square error (RMSE) is taken usually as the fitness value in BFO. The outcomes indicate a promising change in precision and intense lessening in time.

Mahmoud [9] had utilized bacterial searching focused by molecule swarm advancement (BF PSO or essentially BSO) hybridized along with Nelder–Mead (NM) calculation to bow tie type of receiving antenna used for RFID receivers. The BSO–NM calculation is coordinated with the MoM to improve the radio wire. The calculation enhances reception apparatus parameters that is, neck width, stature, and flare edge to make it thunderous at craved frequencies. The coefficient of reflection has been taken as the cost capacity to be minimized. It is found that BFO–NM calculation delivered comes about superior to those created by an individual BFO or BSO.

Chen et al. [12] have presented versatile bacterial scavenging advancement (ABFO) by changing the length of swim step powerfully amid execution of the calculation. Two models ABFO 0 and ABFO 1 have been accounted for. In ABFO 0, the advancement procedure is separated into two stages. The calculation begins with the investigation stage, in which a substantial swim length is doled out to every one of the microscopic organisms. The bigger swim length licenses them to investigate the entire space quickly to find a worldwide ideal and abstain from being caught in neighborhood optima. These places of worldwide ideal are a contribution to the abuse stage. In the misuse stage, the microscopic organisms join the assets found in the past stage and are doled out littler swim step length, so they can better use of the assets. In ABFO 1, every bacterium separately performs an engaged and more profound misuse of promising locales and more extensive investigation of different districts of inquiry space. At the point when a bacterium enters in a supplement wealthier district with higher wellness swim length reductions to have better misuse of the assets. When it enters in a sustenance depleted area, swim length increments to have a quicker investigation. Recreation comes from demonstrating that both ABFOs are certainly superior to the first BFO.

BFO has been utilized as a picture upgrade device in [2]. The paper presents utilization of BFO with a versatile middle channel to enhance crest flag to commotion proportion of a very ruined picture without a unique picture. In first stage, the picture adulterated with salt and pepper commotion of fluctuated thickness is connected to the versatile middle channel. At that point, in second stage both boisterous picture and versatile middle channel yield picture that are gone through BFO to minimize the mistake because of contrasts in separated picture and uproarious picture. Number of microorganisms has been taken same as the quantity of pixels in the picture. The mean square blunder between boisterous picture and separated picture has been taken as cost capacity. Unique picture is reproduced from a picture adulterated with commotion as high as 90%. This quality and precision have been accomplished with concurrent diminishment in computational time.

In [13] Sastri et al. have presented a speed regulated bacterial scavenging improvement system (VMBFO). The velocity modulated BFO (VMBFO) has been gotten from hybridization of BFO and PSO to decrease the joining time. Bigger populace in BFO requires more union time. In any case, if less populace is taken, they might be lacking to investigate the whole inquiry space and there is dependably a probability of being caught in the nearby minima. Taking this issue, the creators took least number of microorganisms and characterized shape before the beginning of the improvement process, and hence staying away from the arbitrary hunt. In the calculation, at first, the whole microscopic organisms are utilized as a part of PSO and are dealt with as particles. After application of PSO, each one of the particles is reassigned as microbes and is permitted to seek haphazardly the correct areas by the method for BFO. The calculation has been utilized to compute the thunderous recurrence of rectangular small-scale strip fix reception apparatus.

Solved Questions

1. Explain the concept of self-adaptation in BFOA and show how it relates to the performance of other algorithms.

   To significantly improve the performance of the original algorithm, self-adaptation [14], a search strategy is employed, which is adaptive in nature. To balance the exploration/exploitation trade-off and to dynamically adjust the run-length unit parameter during evolution, it is achieved by enabling SA–BFO.

   Basic benchmark functions are used in order to compare the performance of algorithms, such as Rosenbrock function, sphere function, Rastrigin function, and Griewank function are used. The experiments we perform run 30 times for each algorithm on each benchmark function and the maximum generation is set at 1000.

   Following are the parameters of the self-adaptive algorithm.

2. Compare the performance of PSO with the variants of BFO algorithm. Can they be hybridized?

Solution

Using PSO and BFOA algorithm, [9] comes up with a hybrid approach for optimizing high-dimensional functions and multimodal functions. The devised algorithm is then used to compare with the originally proposed algorithm.

Following are the operating conditions used for conducting the experiments:

- The mutation step size is set proportional to the maximum acceptable velocity.
- Population size of 40 bacteria is taken.
- To make a fair comparison, the competitor algorithms are initialized using the same random number generator seed.
- We take the values as $C1 = C2 = 1.494$, $N_{re} = 4$, $\omega = 0.8$, $N_c = 50$. This set of parameters is same for all the algorithms (Figures 15.2 and 15.3) [9].

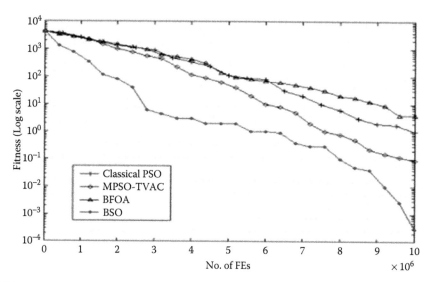

FIGURE 15.2
Progress toward the optima for Ranstrigin function.

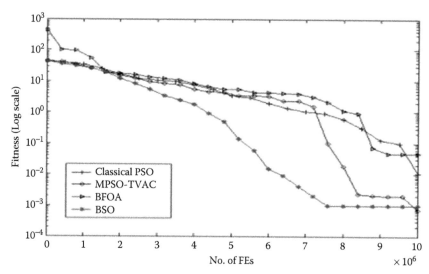

FIGURE 15.3
Progress toward the optima for Ackley function.

Unsolved Questions

1. Highlight the usefulness of the BFO algorithm.
2. Discuss the limitation of the BFO algorithm.
3. Write a short note on the biological inspiration behind the BFO algorithm.
4. What is the physical significance of swimming length?
5. How does elimination–dispersal probability affect the convergence of the algorithm?
6. How is swarming implemented in the BFO algorithm?
7. What factors influence the usability of BFO algorithms in electromagnetic theory particularly antenna problems? Discuss in detail.
8. Discuss the impact of chemotactic parameter on the standard BFO algorithm.
9. What elements make a BFO algorithm adaptive in nature?
10. How can BFO be used along with other soft computing algorithms? Discuss giving concrete evidence.

References

1. K. R. Mahmoud, Design optimization of a bow-tie antenna for 2.45 GHz RFID readers using a hybrid BSO-NM algorithm, *Progress in Electromagnetics Research 100*, 2010, 105–117.
2. K. M. Bakwad, S. S. Pattnaik, B. S. Sohi, S. Devi, B. K. Panigrahi and S. V. Gollapudi, Bacterial foraging optimization technique cascaded with adaptive filter to enhance peak signal to noise ratio from single image, *IETE Journal of Research*, 2009, 55 (4), 173–179.
3. S. Das, A. Bishwas, S. Dasgupta and A. Abhraham, Bacterial foraging optimization algorithm: Theoretical foundation, analysis and its applications, *Foundations of Computational Intelligence*, 2009, 3, 23–55.
4. W. Liu, H. Chen, H. Chen and M. Chen, RFID network scheduling using an adaptive bacteria l foraging algorithm, *Journal of Computer Information Systems (JCIS)*, 2011, 7 (4), 1238–1245.
5. Y. Zhang, L. Wu and S. Wang, Bacterial foraging optimization based neural network for short term load forecasting, *Journal of Colloid and Interface Science*, 2010, 6 (7), 2099–2105.
6. B. B. Mangraj, I. S. Misra and A. K. Barisal, Optimizing included angle of symmetrical V-dipoles for higher directivity using bacteria foraging optimization algorithm, *Progress in Electromagnetics Research B*, 2008, 3, 295–314.
7. T. Datta, I. S. Misra, B. B. Mangraj and S. Imtiaj, Improved adaptive bacteria foraging algorithm in optimization of antenna array for faster convergence, *Progress in Electromagnetics Research C*, 2008, 1, 143–157.
8. G. S. V. R. Sastri, S. S. Pattnaik, O. P. Bajpai, S. Devi, C. V. Sagar, P. K. Patra and K. M. Bakwad, Bacterial foraging optimization technique to calculate resonant frequency of rectangular microstrip antenna, *International Journal of RF and Microwave Computer-Aided Engineering*, 2008, 18, 383–388.

9. A. Biswas, S. Dasgupta, S. Das and A. Abraham. Synergy of PSO and bacterial foraging optimization—A comparative study on numerical benchmarks. In *Innovations in Hybrid Intelligent Systems*, Springer, Berlin, Heidelberg, 2007, 255–263.

10. W. M. Korani, H. T. Dorrah and H. M. Emara. Bacterial foraging oriented by particle swarm optimization strategy for PID tuning. In *Computational Intelligence in Robotics and Automation (CIRA)*, 2009, IEEE International Symposium on 2009, December 15, 445–450.

11. G. S. V. R. Sastri, S. S. Pattnaik, O. P. Bajpai, S. Devi, K. M. Bakwad and P. K. Patra, Intelligent bacterial foraging optimization technique to calculate resonant frequency of RMA, *International Journal of Microwave and Optical Technology*, 2009, 4 (2), 67–75.

12. H. Chen, Y. Zu and K. Hu, Adaptive bacterial foraging optimization, *Abstracts and Applied Analysis*, 2011, 2011, 1–27.

13. G. S. V. R. Sastri, S. S. Pattnaik, O. P. Bajpai, S. Devi and K. M. Bakwad, Velocity modulated bacterial foraging optimization technique (VMBFO), *Applied Soft Computing Journal*, 2011, 11 (1), 154–165.

14. H. Chen, Y. Zhu and K. Hu, Self-adaptation in bacterial foraging optimization algorithm, *3rd International Conference on Intelligent System and Knowledge Engineering*, Xiamen, China, November 17–19, 2008.

16

Flower Pollination Algorithm

16.1 Introduction

Real-world design problems found in the engineering domain are usually multiobjective problems, that is, there is more than one objective that needs to be accomplished at the same time. The trouble lies in the fact that in most of the cases, these objectives go against one another. The solution in these cases is to come to a point where a perfect/optimal balance could be struck between these objectives. In a way, one objective needs to be traded off to get to some part of another and that is where the devil lies. Another option that researchers and problem-solvers go for is approximations and consequent ranking of possible solutions. This is done so that the best ranked solution could be adopted for a given application. This will also help the decision-makers to take smart and proactive decisions given in a scenario. As can be guessed very easily, multiobjective problems have additional challenges such as an increased time complexity, in homogeneity and high dimensionality. To add to this, there is no guarantee that the obtained solution or the chosen rank system will actually help the cause. The Pareto front curves usually form higher order plots such as a surface and so on and this makes the situation even more challenging. A number of nature-inspired optimization algorithms have surfaced in the recent past and their performance has been well documented clearly representing the areas of their utility, their pros as well as cons. Such algorithms include genetic algorithms (GAs) [1], particle swarm optimization (PSO), cuckoo search (CS), and so on, as discussed in the previous chapters. In Sections 16.2 through 16.5, we will be talking about a new algorithm called flower pollination algorithm, a nature-inspired algorithm motivated by how pollination takes place in flowers. We will also be discussing its variants, hybridization with other contemporary algorithms, scope of application as well as applications.

16.2 Flower Pollination

16.2.1 Pollination

The union of gametes is known to bring about reproduction in plants. The pollen grains that are produced by the male gametes and the ovules, which are borne by the male gametes are responsible for this phenomenon. For reproduction to actually take place, it is mandatory that the pollen has to be transferred to the stigma so that union could take place. Pollination thus can effectively be defined as the process of transfer and consequent deposition of pollen grains into the stigma of the flower. For this transfer to actually take place, an agent is required. This agent facilitates the motion of pollen grains that would otherwise have been very restricted [2–5].

Pollination can be divided into the following classes:

- Self-pollination
- Cross-pollination

16.2.2 Self-Pollination

This is an intraplant behavior where the pollen of one flower either pollinates the same flower or the flower present on the same plant. This intra behavior makes it *self*. However, this can take place in only those plants whose flowers contain both the male as well as the female gametes (Figure 16.1).

16.2.3 Cross-Pollination

This, in contrast to self-pollination is an interplant phenomenon. In this case, the pollen grains are transferred from the flower of one plant to the flower of another plant. Various agents such as insects, birds, animals, water, and so on aid the process. The process is called abiotic when the agents are nonliving; for example, wind and so on and if living agents such as insects, bees, beetles, and so on are involved, then the process becomes biotic pollination. Insect pollination can easily be observed for the flowers that have colored petals and a strong odor. The insects in these circumstances are drawn toward these flowers due to abundance of nectar, edible pollens, and so on. When the insect sits on a flower, the pollen grains stick to their body and when the same insect happens to sit on another flower, the pollen grains get dropped.

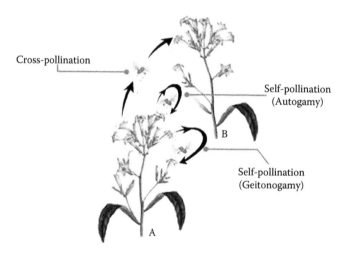

FIGURE 16.1
Types of pollination. (http://biology.tutorvista.com/plant-kingdom/types-of-pollination.html)

16.3 Characteristics of Flower Pollination

It is no surprise that there are millions of flowering plants that the nature contains within its repositories. Though various research works have been done, yet, the dominance of flowering plants in the ecosystem remains a mystery. These plants have always evolved and substantially over the past 120 million years. The natural processes that allow a plant to bear a flower is a wonderful stroke of nature and its beauty. It must be no surprise that certain insects and flowers have coevolved into the so-called partnerships in which certain insects have an affinity for specific flowers. The majority of pollination processes remain biotic in nature, whereas there does exist a small fraction that does not need any pollinator as such. It must also be kept in mind even the pollinators can be highly diverse; for example, it is estimated that there are about 200,000 varieties of pollinators present in nature. These pollinators have specific properties, characteristics, affinity as well as abhorrence to certain plants. A very important as well as useful class of pollinators is honey bees; their usefulness is reached to a point that they are now believed to have developed *flower constancy*. Researchers believe that such a flower constancy may have evolutionary advantages as this would maximize the pollen grain transfer to some specific flowers only. This fact will, in turn, maximize the reproduction of the same flower species. This constancy does also serve the pollinators as they can get a guarantee that nectar amount will be available to them with limited memory and minimum cost of learning. In light of this event, the pollinators do not have to roam about searching for the best flowers; they will tend to move straight toward flowers that are amenable to them. Some survey has also concluded that the flower constancy can represent the increment step in order to refer to the difference/similarity between the two flowers. Furthermore flower constancy can be considered as an increment step using the similarity or difference of two flowers. The flower pollination process is also governed by the Darwin's theory of *survival of the fittest*. This ensures that on reproduction, the plants *born* are the fittest. This in a way optimizes the plants. Thus, in this section, we have discussed about the motivation behind the development of such an algorithm.

16.4 Flower Pollination Algorithm

Xin-She Yang developed the flower pollination algorithm; this algorithm as stated earlier is inspired by the pollination phenomenon that takes place in flowers. This algorithm was also extended to solve multiobjective problems. To understand the algorithm in simple terms, following points need to be kept in mind:

- Global pollination can be achieved by biotic as well as cross-pollination and the pollinator (the agents) is required to obey the levy flight.
- Similarly, in order to implement local pollination, abiotic and self-pollination are preferred.

- Different insects have different flower constancy; this is manifested as a reproduction probability. This probability term remains proportional to the similarity that the two flowers have with one another.
- A term, switch probability, is defined that will enable the algorithm to make a switch from local to global pollination and vice versa. However, given a choice, local pollination is always preferred.

The above-mentioned rules need to be kept in mind to ensure that the algorithm is modeled correctly. For instance, to accommodate the fact that flower gametes are carried by pollinators such as insects and that pollen can travel over long distances can be mathematically put as

$$x_i^{t+1} = x_i^t + \gamma L(\lambda)\left(g^* - x_i^t\right) \tag{16.1}$$

Here, x_i^t is the pollen/solution vector x_i in iteration t, and g^* is the best solution found in the current iteration. The scaling factor γ controls the step size. $L(\lambda)$ is a parameter representing levy-flights-based step size. This factor encapsulates the strength of pollination. Levy flight is used to mimic the long distance flight of insects (it is a well-known fact that the insects can travel/fly long distances).

Though flower pollination activities can occur at all scales, both local and global, adjacent flower patches or flowers in the not-so-far-away neighborhood are more likely to be pollinated by the local flower pollen than those faraway. In order to imitate this, we can effectively use the switch probability or the proximity probability p to switch between the common global pollination to intensive local pollination. To begin with, we can use a naive value of $p = 0.5$ as an initial value. A preliminary parametric showed that $p = 0.8$ might work better for most applications [6].

The basic steps of flower pollination algorithm (FPA) can be summarized as the pseudocode shown in Figure 16.2 [3].

The algorithm goes as follows:

- Initialize a population of F flowers/pollen gametes. Each flower represents a solution, which in the case of combinatorial problems, is a permutation of n nodes.
- Find the best solution g in the initial population.
- Define a switch probability $p \in [0,1]$.
- Generate a random number—rand.
  - If rand $< p$, do global pollination by modifying the flower according to the global best solution.
  - Else, do local pollination by modifying the flower sequence itself, parts by parts, and keep the best one among all the generated sequences.
  - If new solutions are better than the original ones, update them in the population.
- Repeat step 4 for each flower F.
- Find the current best solution g.
- Repeat step 4 to 6 for T iterations.
- The global best solution at the end of T iterations is the best one as generated by FPA.

Flower Pollination Algorithm

1. Define Objective Function $f(x)$, $x = (x_1, x_2,...,x_d)$
2. Initialize a population of n flowers/pollen gametes with random solutions
3. Find the best solution B in the initial population
4. Define a switch probability p [0,1]
5. Define a stopping criterion (either a fixed number of generations/iterations or accuracy)
6. **while** ($t <$ MaxGeneration)
7. **for** $i = 1$: n (all n flowers in the population)
8. **if** (rand $< p$)
9. Draw a (d-dimensional) step vector L which obeys a Lévy distribution
10. Global pollination via $X_i^{t+1} = X_i^t + L(B - X_i^t)$
11. **Else**
12. Draw U from a uniform distribution in [0,1]
13. Do local pollination via $X_i^{t+1} = X_i^t + U(X_j^t - X_k^t)$
14. **End if**
15. Evaluate new solutions
16. If new solutions are better, update them in the population
17. **End for**
18. Find the current best solution B
19. **End while**
20. Output the best solution found

FIGURE 16.2
Pseudocode for the traditional algorithm.

16.5 Multiobjective Flower Pollination Algorithm

The multiobjective flower pollination algorithm (MFPA) is a multiobjective extension of the flower pollination algorithm (single-objective case). This algorithm was proposed by Yang with a view of solving multiobjective problems as well. The obvious difference between single-objective flower pollination algorithm (SFPA) and MFPA is that the former lacks effective search capability as it is restricted to single-objective problems only.

The latter, on the other hand, can solve multiobjective optimization problems [7] as well. A number of varied approaches have been proposed to transform a multiobjective problem into a single-objective problem. A simple such transformation is the use of weighted sum to integrate multiple objectives into a *composite single objective*. This can be modeled as

$$f = \sum_{i=1}^{m} w_i f_i, \quad \sum_{i=1}^{m} w_i = 1, \quad w_i > 0 \tag{16.2}$$

In the above-mentioned equation, m represents the number of objectives, whereas w_i represent nonnegative weights. The random weights w_i are usually drawn from a uniform distribution and are used to get the Pareto front with optimal solutions.

16.6 Variants of Flower Pollination Algorithm

16.6.1 Modified Flower Pollination Algorithm for Global Optimization

The proposed MFPA [8] is described in Figures 16.3 and 16.4 [4] in the form of pseudo-code as well as a flowchart diagram. This makes use of a clonal property as derived from the clonal election principle. To make sure that all words are known to the readers, a separate table (Table 16.1) [4] has been drawn that defines each and every term used in our algorithm. The table also contains the equivalent nomenclature for these terms in other optimization algorithms. The results when analyzed revealed that when the solutions are generated from random walks (as in the local pollination), the algorithm converges faster than the levy flight. This is why levy flights were replaced by random walks. These walks are drawn out of a random distribution bounded by [0,1]. Random walks are drawn from random uniform distribution in [0,1]. In this modification, the high-affinity solutions are cloned in direct proportion to their affinity; it is only then that the local pollination is performed.

Local pollination is modified with the introduction of a step-size scaling factor γ2. A preliminary parametric study showed that γ2 = 3 is an ideal choice for almost all the test cases.

Here, the MFP algorithm selects the 14 best solutions from the population (P_{op}) in iteration t. The solution obtained is cloned in direct proportion to its fitness value. The parameter values of all the five algorithms are highlighted in the Table 16.2. Used cloning value array elements generally do not change; however, it can assume an adaptive tendency in relation to the quality of the underlined solutions. On account of this, the exploitation rate remains high and necessary steps are required to ensure that the solution does not fall in the local minima.

The modified algorithm is validated; for this, its performance is compared with the known algorithms such as FPA, BAT, and so on. For this comparison, 23 standard benchmark functions were used.

The results when tabulated in the Table 16.3 highlighted the fact that this modified algorithm works very well with all the benchmark functions chosen. In some cases, it was also able to surpass the five algorithms (this took place in 12 test cases). As such, it is very safe to say that modified FPA is better than the other five algorithms used for comparison. The value of significance level *alpha* was taken to be 0.05 in all the experiments.

FPA has also been hybridized using the clonal selection algorithm. Following this, the proposed hybridized algorithm was validated using benchmark functions. Some of the benchmark functions used were multidimensional. Again, the hybridized algorithm surpassed the above-mentioned five optimization algorithms.

1. -Minimize or maximize the objective function $f(x)$, $x = (x_1, x_2, ..., x_d)$
2. -Create a random initial population **pop** of size **n**
3. -Identify g^* which is the best solution in pop
4. -Identify $P \in [0,1]$ which is a switch probability between global and local pollination
5. **While** (gen < MaxGenerationsNum)
6. **If rand** > p,   // perform global pollination
7.   **For** each X_i in **Pop**
8.   -Draw a (d-dimensional) step vector L which obeys a levy distribution
9.   -Global pollination via $X_i^{t+1} = X_i^t + \gamma_1 L(g^* - X_i^t)$
10.   **End For**
11. **Else**
12.   -Select best **m** solutions from **Pop** to form **ClonesPop** population.
13.   -Clone solutions in **ClonesPop** proportional to affinity.
14.   **For** each solution in **ClonesPop**
15.   -Draw ϵ from a uniform distribution in [0,1]
16.   -Randomly choose j and k from **Pop**
17.   -Perform local pollination via $X_i^{t+1} = X_i^t + \gamma_2 \epsilon (X_j^t - X_k^t)$
18.   **End For**
19. **End if**
20.   -Select best solutions from **Pop** and **ClonesPop** to form **NewPop** population
21.   -Replace **Pop** by **NewPop**
22.   -Find the current best solution g^*
23.   If g^* doesn't change, for successive 100 iterations, with a value more than 10^{-6},
24.   keep g^* and replace **Pop** by a new randomly generated one.
25. gen = gen +1
26. **End While**
27. **Print** g^*

FIGURE 16.3
Pseudocode depicting the modified flower pollination algorithm.

Given these developments, it can very well be established that the modified FPA can also be extended to discrete applications so that combinatorial optimizations problems such as the traveling salesman problem (TSP) could be solved. The future scope foresees the use of this algorithm for other engineering applications [9] such as power systems, segmentation, quantum theory, and so on.

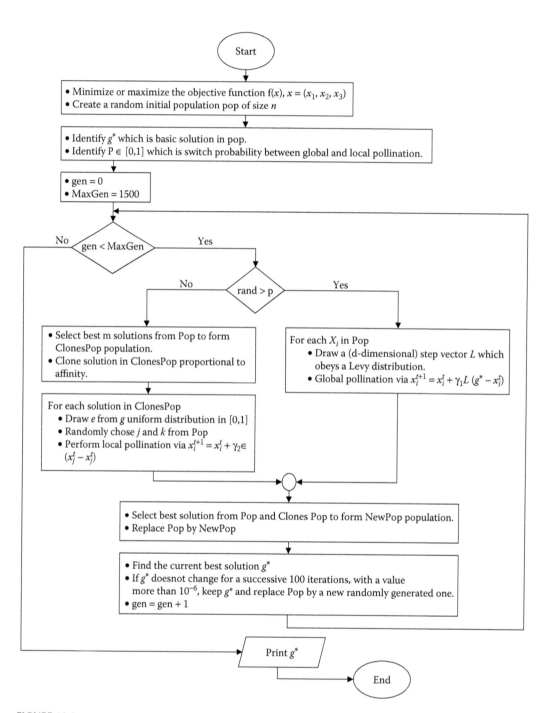

FIGURE 16.4
Flowchart depicting modified flower pollination algorithm.

TABLE 16.1

The Main Terminology of FPA and CSA and Their Equivalence in Traditional Search Algorithms

Terminology from FPA and CSA	Equivalent Terminology
Flower/Pollen	Solution vector
Population of flowers or pollens	Population
Local pollination	Local search
Global pollination	Global sector
Antibody	Solution vector
Antigen	Problem to be solved
Hypermutation	Mutation
Affinity	Fitness
Metadynamics	The process that increases the population diversity
Repertoire	Population

TABLE 16.2

The Main Parameter Setting of Algorithms Used in Experimental Results [4]

Algorithm	Parameters Setting
FPA	$n = 50$, switch probability $P = 0.8$, $\gamma = 0.01$, for Levy flight $\lambda = 1.5$
MFPA	$n = 50$, switch probability $P = 0.8$, $\gamma_1 = 1$, $\gamma_2 = 3$ for Levy flight $\lambda = 1.5$, cloning array $= [9,8,7,6,5,4,3,2,1,1,1,1,1,1]$
GA	$n = 50$, cross over $= 0.8$, mutation function is Gaussian with values: Scale $= 1$ and Shrink $= 1$
BAT	$n = 50$, loudness $= 0.5$, pulse rate $= 0.5$, minimum frequency $= 0$, maximum frequency $= 2$
FF	$n = 50$, randomness value (alpha $= 0.25$), minimum value attractiveness of a firefly (beta $= 0.2$), absorption coefficient (gamma $= 1$), $\gamma = 0.01$, for Levy flight $\lambda = 1.5$
SA	Annealing function: fast annealing, reannealing interval $= 100$, initial temperature $= 100$

TABLE 16.3

Average Ranking of Algorithms by Test for 23 Functions

Algorithm	Mean Rank
FPA	2.30
MFPA	**1.70**
BAT	4.41
FF	3.57
GA	3.89
SA	5.13

Source: Nabil, E., *Expert Sys. Appli.*, 57, 192–203, 2016.

16.6.2 Elite Opposition-Based Flower Pollination Algorithm

Low multidimensional unimodal optimization problems can easily be solved using the traditional FPA. However, for dealing with high-dimensional and multimodal optimization problems, the solutions obtained from basic algorithm are not sufficient [10].

16.6.2.1 Global Elite Opposition-Based Learning Strategy

Traditional algorithm uses the concept of levy flight to conduct global search. The levy flight parameter is estimated by the levy distribution. One thing to keep in mind is that since levy distribution represents a stochastic process, the probability of getting a fitter solution remains remote. Elite opposition-based learning is a novel technique that has recently been added in the domain of intelligent computation. The governing principle of this strategy is that while searching for a better solution, also compute its opposite solution. Toward the end of the iteration, the best solution and its opposite are compared and the fitter between the two makes it to the next iteration.

The above-mentioned strategy expands the searching space of the algorithm and this ensures increased diversity of the population. Thus, in conclusion, the strategy enhances the optimization capability of the algorithm.

16.6.2.2 Local Self-Adaptive Greedy Strategy

It is a well-established fact that standard FPA uses differential evolution algorithm [11,12] to perform local search and needless to say, the searching ability of the DE algorithm is not very impressive especially for high-dimensional optimization problems. This calls for a local self-adaptive greedy strategy to enhance the exploitation capability of the algorithm.

When this strategy is put in perspective, the adjustments are larger in the initial stages of the algorithm thereby promoting exploration so that the solution jumps out of local minima. However, with increasing iterations, the adjustments start to decrease in value facilitating in the process the exploitation capability of the algorithm.

16.6.2.3 Dynamic Switching Probility Strategy

In FPA, *switch probability* dictates the degree of local and global search. This term is represented by the variable p and is a constant. The rationale behind the use of switch probability is the fact that a reasonable algorithm is expected to perform more of exploration in the initial stages and exploitation in the later stages.

16.7 Application of Flower Pollination Algorithm

16.7.1 The Single-Objective Flower Pollination Algorithm

Yang [3] proposed SFPA; the motivating factor was the fact that limited number of nature-inspired algorithms were available. Thus, in light of this event, the introduction of FPA did to some extent fill the vacuum present. In his work, Yang implemented the SFPA and tested it on standard benchmark functions. He concluded from the results that his algorithm was far better than PSO [13] and GA. This algorithm was also implemented in wireless sensor networks (WSN) [14]. The objective was to select cluster heads and distribute the cluster nodes in a way that the setup overcomes the limitations of LEACH protocol. When the algorithm was implemented in real time, the results suggested the usefulness of the FPA. As a consequence, the lifetime of the network improved. The algorithm has also been used in the optimization of retinal vessel segmentation [15]. The process of optimization was carried out in two stages. In the first stage, the algorithm searches greedily the optimal vessels present in the retina. Following this, the cluster centers obtained are used to perform local searches. When tested, the algorithm proved promising in terms of faster convergence to optimal solutions. In yet another work, FPA was compared with the traditional Bat algorithm. The benchmarks functions used were unimodal as well as multimodal. High-dimensional continuous functions were also considered. As expected by the authors, the performance of the FPA was better. Following this, the authors in [16] compared the performance of the SFPA with that of PSO on continuous benchmark functions. The results did not disappoint the authors as the performance of SFPA again outshined the of PSO algorithm.

16.7.2 The Multiobjective Flower Pollination Algorithm

The authors in [17] extended the SFPA to MFPA with the sole objective of solving multiobjective problems [18]. The performance of the modified algorithm was evaluated on standard benchmark functions and structural design. The performance MFPA was also compared with other algorithms such as VEGA, NSGA-II, and multiobjective differential evolution (MODE) [19], GA, and so on [20–23].

16.7.3 The Hybrid Flower Pollination Algorithm

Hybridization generally involves the use of more than one algorithm to solve our problem more efficiently. One thing to take notice of is the fact that not all algorithms can be mixed with a given soft computing algorithm. As the analysis of this is beyond the scope of this book, so we are ignoring it for time being. One obvious advantage that hybridization brings with itself is the fact that weakness of the algorithm gets compensated in one way. This increases the potential for getting more effective solutions in reduced number of iterations. [24] proposes a hybrid version of SFPA and FPCHS. The primary objective of hybridization was to improve on the accuracy of the solutions. The performance of hybridization was then evaluated on Sudoku puzzles.

Results obtained from experimentation asserted that the hybridized algorithm was indeed better in providing better convergence to the optimal solution. On the same grounds, SFPA was also hybridized with PSO algorithm [25] to improve the performance of the former. To add to this series of hybridization, the [26] came up with ESSFPA to get a balance between exploration and exploitation capabilities of the algorithm. For this, unimodal as well as multimodal functions were used. The researchers concluded from the results that ESSFPA utilized only one-tenth of computational faculties compared to PSO and yet was able to provide optimal solutions. Another work worth mentioning is the one presented in [27]. Here, a neighborhood searching strategy is introduced to enhance the quality of the solutions. The simulation results obtained when the algorithm was tested on standard benchmark functions suggest an obvious improvement in the performance of the algorithm. The notion of hybridization was also utilized in power systems [28] where SFPA was integrated with PSO to get SFPAPSO. This hybrid version was evaluated on a standard IEEE 30, IEEE 57 bus test systems.

16.8 Conclusion

In this chapter, a number of different applications of the algorithm have been considered. The idea is to provide a brief overview of the same so that new prospects could be discovered and implemented. This will also promote ground research in this algorithm.

There are two classes of FPA: SFPA and MMFPA. Although the former concerns itself with solving single-objective problems, the latter is frequently used for solving multiobjective problems. The review work done in this chapter highlights the fact that this algorithm is useful in a number of areas such as structural and mechanical designs, energy applications, gaming, and so on.

Further, when the results of this algorithm were inspected and then compared with contemporary algorithms such as GA, PSO, and so on the researchers were amazed to see the gain in the performance they were able to get. In addition, as the algorithm is a bit newer and in its infancy, the future scope of this algorithm remains immense. The current analysis also underscores the fact that FPA has got a vast potential in areas such as petroleum industry, carbon dioxide emission, data mining, and so on.

Solved Questions

1. Use FPA to solve TSP problem.

Solution

We use the following rules to apply the FPA algorithm here:

   a. Biotic and cross-pollination are considered as global pollination mechanisms.

   b. Similarly, we consider abiotic and self-pollination as local pollination mechanism.

   c. Switch probability is used to control the degree of local and global pollination. This factor remains in [0,1]. In majority of circumstances, local pollination dictates the probability value (due to presence of wind).

The algorithm is coded in C++ and standard datasets have been taken. The following are the results obtained as presented in Tables 16.4 and 16.5.

2. Calculate flower constancy given time $(t + 1)$ given that the solution vector at time (t) is 23 and levy-flights-based step size is 0.44. Assume the scaling factor to be 2.4 and the best solution to be 44.

TABLE 16.4

The Code is Tested on Berlin52 Dataset with 52 Cities

Given Optimal	Mean over 5 Run	Worst Obtained	Best Obtained	Error
7498	9325	9987	8876	1827

TABLE 16.5

Table Results Obtained Using QAP Dataset

Dataset	Number of Iterations	Given Best	Mean over 5 Runs	Best Obtained	Worst Obtained	Error
Nug12	20	578	588	578	600	10
Esc16i	10	14	14	14	14	0
Kra32	50	88,700	91,776	90,900	92,850	3,076

Solution

Using,

$$x_i^{t+1} = x_i^t + \gamma L(\lambda)\left(g^* - x_i^t\right) \tag{16.1}$$

We have,

$$x_i^{t+1} = 23 + 2.4 * 0.44 * (44 - 23)$$

$$x_i^{t+1} = 45.176$$

Unsolved Questions

1. Differentiate between self- and cross-pollination. Also comment on the biological inspiration behind FPA.
2. How does pollination take place biologically?
3. Highlight the distinguished characteristics of the FPA.
4. Comment on the usefulness of FPA.
5. Highlight the limitations of FPA algorithm.
6. How is global pollination different from local pollination?
7. What is the physical significance of switch probability? Does it affect the convergence of the algorithm? Explain.
8. How is metadynamics implemented in FPA algorithm?
9. Write a short note on local self-adaptive greedy strategy.
10. Discuss the rationale behind elite opposition-based FPA.

References

1. D. E. Goldberg, *Genetic Algorithms in Search, Optimisation and Machine Learning*, Addison Wesley, Reading, MA, 1989.
2. P. Willmer, *Pollination and Floral Ecology*, Princeton University Press, Princeton, NJ, 2011.
3. Flower Pollination by biology.tutorvista.com/animal-kingdom, Pearson, 2005.
4. B. J. Glover, *Understanding Flowers and Flowering: An Integrated Approach*, Oxford University Press, Oxford, UK, 2007.
5. M. Walker, How flowers conquered the world, *BBC Earth News*, http://news.bbc.co.uk/earth/hi/earth news/newsid 8143000/8143095.stm, 2009.
6. X. S. Yang, Flower pollination algorithm for global optimization, In Durand-Lose, J. and Jonoska, N. (Eds.), *Unconventional Computation and Natural Computation, Lecture Notes in Computer Science*, Vol. 7445, pp. 240–249. Springer-Verlag, Berlin, Germany, 2012.
7. C. A. Floudas, P. M. Pardalos, C. S. Adjiman, W. R. Esposito, Z. H. Gumus, S. T. Harding, J. L. Klepeis, C. A. Meyer and C. A. Scheiger, *Handbook of Test Problems in Local and Global Optimization*, Springer, Dordrecht, the Netherlands, 1999.
8. E. Nabil, A modified flower pollination algorithm for global optimization, *Expert Systems with Applications*, 2016, 57, 192–203.
9. H. Chiromaa, N. L. M. Shuibb, S. A. Muazc, A. I. Abubakard, L. B. Ilae and J. Z. Maitamaf, A review of the applications of bio-inspired flower pollination algorithm, *Procedia Computer Science*, 2015, 62, 435–441.
10. Y. Zhou, R. Wang and Q. Luo, Elite opposition-based flower pollination algorithm, *Neurocomputing*, 2016, 188, 294–310.
11. H. A. Abbass and R. Sarker, The pareto differential evolution algorithm, *International Journal on Artificial Intelligence Tools*, 2002, 11 (4), 531–552.
12. A. H. Gandomi, X. S. Yang, S. Talatahari and S. Deb, Coupled eagle strategy and differential evolution for unconstrained and constrained global optimization, *Computers & Mathematics with Applications*, 2012, 63 (1), 191–200.
13. L. C. Cagnina, S. C. Esquivel and C. A. Coello, Solving engineering optimization problems with the simple constrained particle swarm optimizer, *Informatica*, 2008, 32 (2), 319–326.
14. M. Sharawi, E. Emary, I. A. Saroit and E. H. Mahdy, Flower pollination optimization algorithm for wireless sensor network lifetime global optimization, *International Journal of Soft Computing and Engineering*, 2014, 4, 54–59.
15. E. Emary, H. M. Zawbaa, A. E. Hassanien, M. F. Tolba and V. Snasel, Retinal vessel segmentation based on flower pollination search algorithm, *Proceedings of the Fifth International Conference on Innovations in Bio-Inspired Computing and Applications IBICA*, 2014, Springer, Cham, pp. 93–100.
16. S. Łukasik and P. A. Kowalski, Study of flower pollination algorithm for continuous optimization, In *Intelligent Systems*, 2015, Springer, Cham, 451–459.
17. X. S. Yang, S. Deb and X. He, Eagle strategy with flower algorithm, *Advances in Computing, Communications and Informatics (ICACCI)*, Mysore, India, August 22–25, 2013.
18. X. S. Yang, M. Karamanoglu and X. He, Flower pollination algorithm: A novel approach for multi-objective optimization, *Engineering Optimization*, 2014, 46, 1222–1237.
19. B. V. Babu and A. M. Gujarathi, Multi-objective differential evolution (MODE) for optimization of supply chain planning and management, *IEEE Congress on Evolutionary Computation (CEC 2007)*, 2007, 2732–2739.
20. C. A. C. Coello, An updated survey of evolutionary multi-objective optimization techniques: State of the art and future trends, *Proceedings of 1999 Congress on Evolutionary Computation (CEC 99)*, 1999, 1, 1–33.
21. K. Deb, Evolutionary algorithms for multi-criterion optimization in engineering design, In K. Miettinen, P. Neittaanmäki, M. M. Mäkelä and J. Périaux (Eds.), *Evolutionary Algorithms in Engineering and Computer Science*, pp. 135–161. Wiley, Hoboken, NJ, 1999.

22. K. Deb, A. Pratap and S. Moitra, Mechanical component design for multiple objectives using elitist non-dominated sorting GA, *Proceedings of the Parallel Problem Solving from Nature VI Conference*, Paris, France, September 18–20, 2000, pp. 859–868.

23. K. Deb, *Multi-Objective Optimization using Evolutionary Algorithms*, John Wiley & Sons, New York, 2001.

24. O. A. Raouf, M. A. Baset and I. E. Henawy, A novel hybrid flower pollination algorithm with chaotic harmony search for solving sudoku puzzles, *International Journal of Modern Education and Computer Science*, 2014, 3, 38–44.

25. O. A. Raouf, A. M. Baset and I. E. Henawy, A new hybrid flower pollination algorithm for solving constrained global optimization problems, *International Journal of Applied*, 2014, 4, 1–13.

26. X. S. Yang, S. Deb and X. He, Eagle strategy with flower algorithm, In *Advances in Computing, Communications and Informatics (ICACCI), 2013 International Conference on 2003*, 2003, 1213–1217.

27. R. Wang and Y. Zhou, Flower pollination algorithm with dimension by dimension improvement, *Mathematical Problems in Engineering*, 2014. doi:10.1155/2014/481791.

28. L. Kanagasabai and B. R. Reddy, Reduction of real power loss by using fusion of flower pollination algorithm with particle swarm optimization, *Journal of the Institute of Industrial Applications Engineers*, 2014, 2, 97–103.

Index

Note: Page numbers followed by f and t refer to figures and tables respectively.

Milton Keynes UK
Ingram Content Group UK Ltd.
UKHW051947071024
449327UK00026B/2205